AF699014

Animal Spatial Cognition:
Behavioural and Brain Approach

Animal Spatial Cognition:
Behavioural and Brain Approach

Catherine Thinus-Blanc

Centre of Research for Cognitive Science, France

World Scientific
Singapore • New Jersey • London • Hong Kong

Published by

World Scientific Publishing Co. Pte. Ltd.

P O Box 128, Farrer Road, Singapore 912805

USA office: Suite 1B, 1060 Main Street, River Edge, NJ 07661

UK office: 57 Shelton Street, Covent Garden, London WC2H 9HE

British Library Cataloguing-in-Publication Data
A catalogue record for this book is available from the British Library.

ANIMAL SPATIAL COGNITION

Copyright © 1996 by World Scientific Publishing Co. Pte. Ltd.

All rights reserved. This book, or parts thereof, may not be reproduced in any form or by any means, electronic or mechanical, including photocopying, recording or any information storage and retrieval system now known or to be invented, without written permission from the Publisher.

For photocopying of material in this volume, please pay a copying fee through the Copyright Clearance Center, Inc., 222 Rosewood Drive, Danvers, MA 01923, USA. In this case permission to photocopy is not required from the publisher.

ISBN 981-02-2818-X

Printed in Singapore.

QL
785
.T45
1996

To Pitt, Emilie
and Jean-Eymeric

PREFACE

When I began to study animal spatial cognition, the theoretical framework was dominated by the still live controversy between the mainstream Cognitivist and Behaviorist schools. The primary concern for the Cognitivist supporters of E. C. Tolman (among whom I numbered) was to demonstrate that spatial maps really do exist in animals. The defenders of Behaviorism, overshooting Hull's subtle statements, tried to account for most spatial behaviors in terms of stimulus-response (S-R) associations. It was against this background that John O'Keefe and Lynn Nadel's book, "The Hippocampus as a Cognitive Map", was published in 1978. Among the many reasons for the success of the book was the fact that within the model that it proposed, the two rival camps became reconciled.

That allowed us to advance beyond the original Cognitivist-Behaviorist controversy. Without doubt, it was the publication of O'Keefe and Nadel's seminal work that paved the way for the fertile multidisciplinary approach that has since ensued. It is a main purpose of this book to review the most noteworthy studies in the field of spatial cognition at behavior and brain levels. Another aim is to draw the reader's attention to the ambiguous meaning of the concept "cognitive maps". Indeed, the cognitive map hypothesis launched by Tolman in 1948 contrasted with the widely accepted notion that animals are not intelligent beings. This hypothesis has strongly influenced, and to some extent biased, the evolution of studies on animal spatial cognition. That is the reason why it is sensible now to examine and interpret recent data in the light of past history. Without such an early hypothesis which endowed animals with the highest levels of cognitive capability, how would research on animal spatial cognition have evolved?

My colleagues and friends have supported me in this enterprise. I have no words, even in French, to express my gratitude to Nigel Foreman who revised the whole manuscript and encouraged me with deep friendship. He improved my awkward English but also greatly improved the form and content of the manuscript. Of course, any flaws that may remain in the manuscript are my responsibility alone. I am also very grateful to Bruno Poucet whose cartesian mind prevented me from making too many digressions. The revisions that he suggested have been the origin of many fruitful and heated discussions. Several other colleagues have provided me with important pieces of information, or have read some parts of the manuscript and have encouraged me. I would like to thank Simon Behamou, Cristina Broglio, Marie-Christine Buhot, Nicole Chapuis, Arianne Etienne, Florence Gaunet, Thom Herrmann, Jacques Ninio, Robert

Muller, Jacques Paillard, Patrick Péruch, Fernando Rodriguez, Cosme Salas, Etienne Save, Françoise Schenk and Wolf Singer.

At this time, I have a special thought for the late Jean Requin, Director of the Center for Research in Cognitive Neuroscience. He possessed the art of heartening failing enthusiasms with convincing arguments drawn from his comprehensive and deep knowledge of the Cognitive Neurosciences.

I am very indebted to Patricia Scardigli who, with care and competence, undertook statistical processing and the preparation of many of the illustrations. The experiments would have not been possible without Henriette Lucchessi whose patience and gentleness succeeded in transforming stubborn rats into cooperating companions. Geneviève Blanchet undertook, with an invaluably good mood, the painstaking collection of documents and typewriting of references. Lastly, without the invaluable help and very high level of competence of Françoise Joubaud, who carried out the camera-ready copy, the completion of this book could not have been achieved. I thank her for her infinite patience and kindness in spite of many last-minute revisions. I would like also to thank the team of the central library of the CNRS. They did their very best to find the most obscure references and articles.

C. Thinus-Blanc
Aix-en-Provence, 12 July 1996

ACKNOWLEDGEMENTS

I would like to thank B.L. McNaughton for permission to use material from his publications.

The following organizations have kindly allowed me to reproduce material:

- Fig. 2.2, reprinted from *Behavioural Processes, 26,* C. Thinus-Blanc, M. Durup, & B. Poucet (1992), pp. 43-57; Figs. 6.7 and 6.8, reprinted from *Behavioural Brain Research, 47,* E. Save, M.-C. Buhot, N. Foreman, & C. Thinus-Blanc (1992), pp. 113-127; Fig. 6.9, reprinted from *Behavioural Brain Research, 37,* B. Poucet & T. Herrmann (1990), pp. 269-280; Figs. 7.2 and 7.3, reprinted from *Behavioural Brain Research,* C. Thinus-Blanc, E. Save, B. Poucet, & N. Foreman, in press; Fig. 7.4 reprinted from *Behavioural Brain Research, 23,* B. Kolb & J. Walkey (1987), pp. 127-145; and Fig. 8.1, reprinted from *Behavioural Brain Research, 41,* B. Poucet (1990), pp. 229-237, with kind permission of Elsevier Science, Sara Burgerhartstraat 25, 1055 KV Amsterdam, The Netherlands.

- Fig. 6.1, reprinted from J. O'Keefe and L. Nadel, *The hippocampus as a cognitive map* (1978), with kind permission of Oxford University Press, Walton Street, Oxford OX2 6DP, U.K.

- Fig. 2.9, reprinted from *Quarterly Journal of Experimental Psychology, 42B,* B. Poucet, B. Bolson, & T. Herrmann (1990), pp. 369-384; Fig. 3.8, reprinted from *Quarterly Journal of Experimental Psychology, 37B,* B. Poucet (1985), pp. 155-179; and Fig. 4.7, reprinted from *Quarterly Journal of Experimental Psychology, 39B,* N. Chapuis & C. Varlet (1987), pp. 49-64, by permission of the Experimental Psychology Society.

- Fig. 5.5, reprinted from *American Psychologist, 34,* D.S. Olton (1979), pp. 583-596; Figs. 5.6, 5.7, and 9.5, reprinted from *Psychological Review, 100,* B. Poucet (1993), pp. 163-182; Figs. 6.5 and 6.6, reprinted from *Behavioral Neuroscience, 106,* E. Save, B. Poucet, N. Foreman, & M.-C. Buhot (1992), pp. 447-456; Fig. 7.5, reprinted from *Behavioral Neuroscience, 110,* E. Save & M. Moghaddam (1996), pp. 74-85; and Figs. 8.2 and 8.4, reprinted from *Behavioral Neuroscience, 109,* S. Granon & B. Poucet (1995), pp. 474-484, with kind permission of the American Psychological Association.

- Figs. 5.2 and 5.7, reprinted from *Psychobiology, 17,* B.L. McNaughton, B. Leonard, & L. Chen (1989), pp. 230-235, by permission of Psychonomic Society, Inc., 1710 Fortview Road, Austin, Texas 78704, U.S.A.

- Fig. 2.6, reprinted from *Hippocampus, 1,* C. Thinus-Blanc, E. Save, B. Poucet, & M.-C. Buhot (1991), pp. 365-372; Fig. 9.2, reprinted from *Hippocampus, 1,* R.U. Muller, J.L. Kubie, & R. Saypoff (1991), pp. 243-246; and Fig. 9.3, reprinted from *Hippocampus, 1,* J. O'Keefe (1991), pp. 230-235, with kind permission of John Wiley and Sons, Inc.

- Fig. 5.4, reprinted from K. Cheng and C.R. Gallistel, in H. Roitblat, T.G. Bever & H.S. Terrace (Eds.), *Animal Cognition* (1984), pp. 409-423, with kind permission of Lawrence Erlbaum Associates, Inc. Hillsdale, NJ 07642, U.S.A.

- Figs. 3.1 and 3.2, reprinted from *Learning and Motivation, 14,* B. Poucet, M.C. Buhot-Averseng, & C. Thinus-Blanc (1983), pp. 140-153; and Figs. 3.5 and 3.6, reprinted from *Learning and Motivation, 14,* D.S. Mazmanian & W.A. Roberts (1983), pp. 123-140, with kind permission of Academic Press, Inc., Orlando, FL 32887-4900, U.S.A.

- Fig. 7.1, reprinted from J.R. Krebs, in B. Kolb & R.C. Tees (Eds.), *The cerebral cortex of the rat* (1990), pp. 459-472; Fig. 8.5, reprinted from P. Dean, in B. Kolb & R.C. Tees (Eds.), *The cerebral cortex of the rat* (1990), pp. 275-308; Figs. 9.1, 10.4 and 10.6, reprinted from C. Thinus-Blanc, in H.L. Rothblat & J.A. Meyer (Eds.), *Comparative approaches to cognitive science* (1995), pp. 241-270; and Fig. 9.4, reprinted from McNaughton, in L. Nadel, L.A. Cooper, P. Culicover, & R.M. Harmish (Eds.), *Neural connections and mental computations* (1989), pp. 285-349, with kind permission of the M.I.T. Press, Cambridge, MA 02142, U.S.A.

- Fig. 3.9, reprinted from B. Poucet and G. Scotto, in P. Ellen & C. Thinus-Blanc (Eds.), *Cognitive processes and spatial orientation in animal and man* (1987), pp. 135-146; and Fig. 10.5, reprinted from F. Schenk, M.C. Grobéty, P. Lavenex and H.P. Lipp, in E. Alleva et al. (Eds.), *Behavioural brain research in naturalistic and semi-naturalistic settings* (1995), pp. 277-300, with kind permission of Martinus Nijhoff Publishers and Kluwer Academic Publishers, 3300 AH Dordrecht, The Netherlands.

CONTENTS

CHAPTER 1

THE HISTORICAL BACKGROUND TO THE STUDIES OF ANIMAL SPATIAL COGNITION: THE CONCEPT OF A COGNITIVE MAP

"The apparent victory of cognitive psychology over behaviourism and the subsequent uncritical proliferation of "cognitive" has involved a cost: a confusion between a school of thought and a process. It has made grey some fundamental distinctions between different processes, obscuring the fact that something like S-R links can be formed, instead of, or in addition to, cognitions." (Toates, 1994a, p. 105)

The survival of many species depends on their ability to use spatial knowledge. Rodents searching for food far from their burrow must escape the shadow of the predator or they will be caught. A pigeon released several hundred miles from its loft must be able to find the right direction home, otherwise it will get lost and die. In this hard and hostile world, animals must know where to find places where food is plentiful and keep in mind their homeward direction in case of approaching danger. This helps us understand why spatial abilities are well developed in many animal species, and have, for some animals such as homing pigeons and bees, become highly sophisticated. Thus it is quite understandable why animals' spatial representations have been debated through the five decades from the infancy of experimental psychology, and why they are still being investigated in many studies on a wide range of species.

As far back as the 1930's, a number of studies were conducted on spatial representations in animals. Many of these experiments were clever but have now fallen into oblivion. They can be found listed in Munn's (1950) "Handbook on Psychological Research on the Rat".

Shepard (1931), for instance, trained rats in the apparatus shown in Fig. 1.1. When at A, the alley was closed at F until the rat reached B. When at E, it had to proceed to F to get a reward. After several trials in this loop, the animals refused to go ahead and stopped obstinately at level F.

In another experiment by the same author (1933), the complex maze shown in Fig. 1.2 was used. After the rat had learned to reach the food box without entering any dead ends, an opening was made at point X. During the first trial following this change, the rat explored the alley b for a short distance once it arrived at segment c, and then continued along the usual runway. However, in the next trial, it chose segment a, which led initially to a blind alley b, and reached the goal by taking a shortcut. This shortcut behavior is nowadays taken as a classic demonstration of the existence of overall spatial representations; that is, the subject proceeds by representing the relative locations of places which can be connected by various pathways (depending on the constraints of the situation). At that time, however, these data were interpreted using terms such as "animal intelligence", "insight", "reasoning", but seldom was reference made explicitly to space, and never to spatial representations. Indeed, the main concern was rather general, concerned with in determining whether animals are "intelligent" beings.

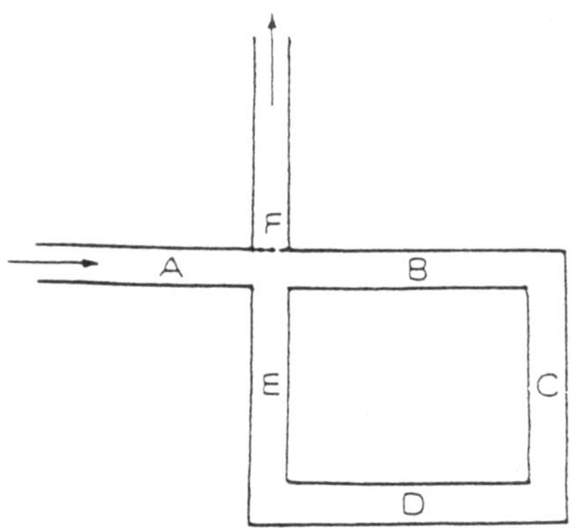

Fig. 1.1. Apparatus used by Shepard in 1931 (after Munn, 1950).

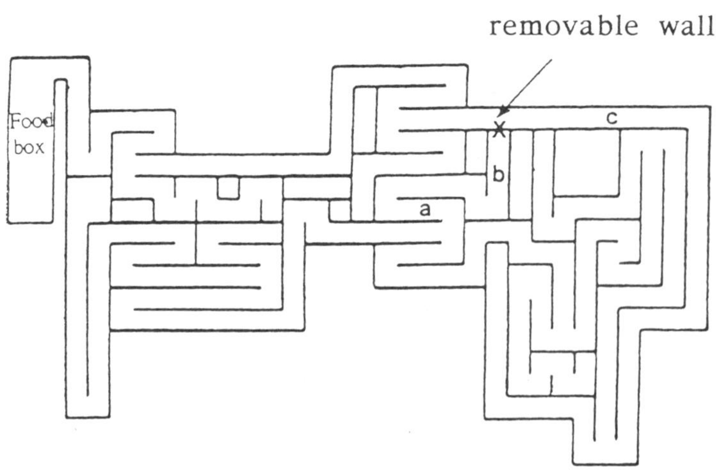

Fig. 1.2. Apparatus used by Shepard in 1933 (after Munn, 1950).

In contrast, after 1948, most of the authors of studies on spatial cognition in animals and humans have referred to the concept of a "cognitive map". This concept was coined by Tolman in his provocative 1948 article "Cognitive Maps in Animals and Man". Since then, the term "cognitive map" has become widely used and is usually taken to refer to a very high level of spatial processing, involving a kind of "bird's eye view" of the environment, which makes it possible to move efficiently between places charted on the map. Many disciplines have taken over this suggestive expression: psychology, of course, both animal and human, as well as ethology, neuroscience, architecture, town planning, geography, ethnology and artificial intelligence. This common usage of the term cognitive map as if it was an objective fact might give the impression that it refers to an entity which has been proved to exist. In fact, the form and the content of the cognitive map are still very puzzling, particularly in the case of animals, and one cannot help wondering how the term became so widely accepted in the first place, since the arguments initially presented by Tolman to support the existence of cognitive maps were really rather weak.

It was recently pointed out by Toates (1994a) that the term "cognitive" is ubiquitous in behavioral science; it has so many different meanings that it does not prove helpful unless we have a clear definition of what is a cognitive process. Toates defines cognitive processes as "... those that involve the assimilation, storage and utilization of knowledge, in a way that is not linked directly to any particular behaviour. Cognition allows extrapolation beyond the sensory input and permits the maximum possible range of varied and flexible behaviour." (Toates, 1994a, p. 102). Such a definition implies, ipso facto, that there is a hierarchy among the mechanisms which control behavior.

The main risk inherent in the wide use of this rather loosely defined term to account for a host of data of various kinds is that these data are liable to be misinterpreted. Not only may behavioral appraisals suffer from this flaw, but also the conclusions drawn from brain studies designed to investigate which brain areas are involved in spatial processing. To be able to understand the specialized role of a given brain structure it is necessary first to analyse as accurately as possible the psychological and behavioral mechanisms that are presumably subserved by this structure.

The general purpose of this book is to examine animal spatial representations, and, correspondingly, the concept of a cognitive map, in the light of the behavioral and brain data which have accumulated since 1948. Do we have more grounds nowadays for speaking about cognitive maps in animals than we had fifty years ago? Is it possible to define exactly their nature and their properties? Do they constitute the sole means of spatial representation or do there exist other high level spatial encoding systems?

In this chapter, I propose to outline Tolman's hypothesis about the existence of cognitive maps in animals and to give an idea of its historical context. This provocative hypothesis marked the beginning of a series of studies focusing specifically on spatial representation in animals. Up to then, spatial processes were thought rather to be ancillary clues for animal intelligence. I will then trace the development of the concept of cognitive maps up to the present day, highlighting the main scientific events in this connection. The general questions raised by the various approaches and theories will be discussed in the light of published data and of the theories put forward since that time. The ultimate purpose of this book is to build up a reasonable picture of animal spatial representation in terms of both the behavioral and the central brain processing involved.

1. The Hull-Tolman Controversy

Before examining Tolman's hypothesis, it is necessary to look at the history of the ideas forming the scientific context for the development of Tolman's own. At the end of the last century, psychological studies were based on the use of the introspective method. This approach gave rise to a reaction in philosophy leading to the emergence of *logical positivism*, and in psychology, to *behaviorism*. The main concern of behaviorism was to imitate the objective methods of physics. The first behaviorists such as Watson and Guthrie aimed at "purifying" Psychology of any form of mentalism in order to transform it into an exact science. The implication was that behavior should just be *observed as a phenomenon*, without making any conjectures about what might be going on within the subject's head which was viewed as a "black box".

Clark Hull was one of the most influential proponents of behaviorism. In a number of books and papers (1930, 1932, 1934a and b, 1943, 1952), he applied his general theory to spatial behavior in animals and man. Complex trajectories were accounted for in terms of stimulus-response (S-R) associations. In spatial mazes such as in Fig. 1.3, many cues (visual, auditory, olfactory, kinesthetic, etc.) can act as signs (stimuli) which indicate to the animal what turns should be made. The rat learns that a particular stimulus is always associated with a specific turn or a specific distance to be run.

For the S-R association to be established, it is necessary that a correct response should be quickly followed by a reward, and an incorrect response by a lack of reward or a punishment In this kind of "response learning" by trial and error, the behavior is thought of as a series of *reactions* driven by environmental stimuli rather than arising spontaneously. Motivation for a reward is crucial, since it means that the S-R associations are strengthened by the "drive-reduction" mechanism. Hull's main concern was to understand how the subject learns to respond to a stimulus; that is, how S-R connections are set up. To make any inferences about what might have been happening in the rat's head would have been too speculative for Hull, so only observable phenomena, namely the physical environment and the subject's actions, were to be taken into account for conclusions to be drawn about the simple relationships or associations between the two entities.

Although Tolman began his career as a relatively orthodox behaviorist, he gradually moved away from this position, and his 1948 article marked the break with the conventional behaviorist school and triggered a long lasting intellectual debate. All the experimental arguments that he developed in support of his cognitive map hypothesis in fact challenged the behaviorists' views, particularly Hull's drive-reduction theory of spatial behavior.

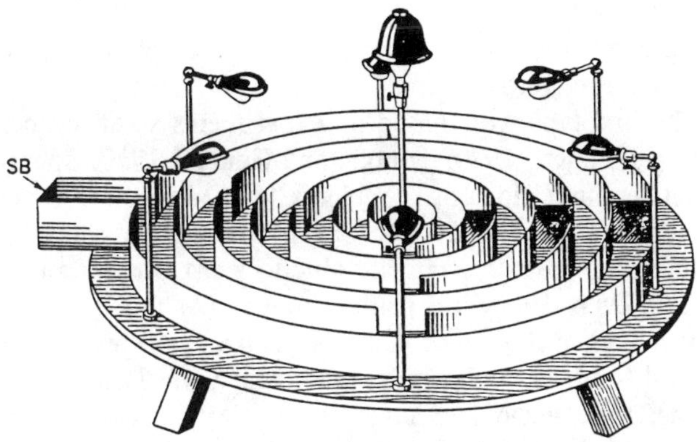

Fig. 1.3. An example of maze used in early research; the Watson circular maze. SB is the starting box; the food chamber is in the center. The path to be followed to reach the food consists in an alternation of right and left turns (reproduced from Munn, 1950).

Tolman's 1948 article was a real bombshell in the behaviorist world and attracted fierce criticism of his general theory of learning, and particularly of his cognitive map hypothesis. He was accused of reverting to the abhorred methods of Introspectionism. Guthrie (1952), for instance, blamed him for leaving the animal "buried deep in thought" and for making no attempt to account for or predict observable behavior. Indeed, as Blancheteau (1969) and Bolles (1975) have pointed out, Tolman overreacted in his forceful criticism of mechanistic models and deliberately declared war on behaviorism, adopting an aggressive and provocative style of writing.

Tolman defined cognitive maps as follows:

"*We believe that in the course of learning, something like a field map of the environment gets established in the rat's brain ... the incoming impulses are usually worked over and elaborated into a tentative cognitive-like map of the environment. And it is this tentative map, indicating routes and paths and environmental relationships, which finally determines what responses, if any, the animal will finally release*." (Tolman, 1948, p. 192).

Since the information is acquired at time x but may not be used until later at time y, Tolman referred to this kind of spatial learning as "latent learning". Tolman's position was based on two main assumptions:

"*First, that latent learning consists not in stimulus-response connections but in the building up in the nervous system of sets which function like cognitive maps, and second, that such cognitive maps may be usefully characterized as varying from a narrow strip variety to a broader comprehensive variety*" (Tolman, 1948, p. 193).

Tolman presented five types of arguments which help us to better understand his definition of cognitive maps.

Those of the first type were based on experiments showing the existence of "latent learning" (Blodgett, 1929; Tolman and Honzik, 1930; Spence and Lipitt, 1946). In latent learning situations, subjects learn about the characteristics of some places even if those characteristics are not relevant to the subject's current motivational state. For instance, rats are able to learn where food and water are located in a maze even if they are neither hungry nor thirsty at the time. What mattered most to Tolman here was that this phenomenon provided good ammunition against Hull's "drive-reduction" model which was based on the idea that a reduction in the drive occurring immediately after a response contributes to learning by strengthening S-R associations.

The arguments of the second type were based on "Vicarious Trials and Errors" (VTE; Geïer, Levine, and Tolman, 1941; Jackson, 1943). VTE involve those behaviors which can be observed when an animal has to make a choice. In this situation, a rat will typically hesitate, looking first to one side, then to the other, etc. VTE is usually observed at the very beginning of discrimination learning or just before the learning criterion has been reached. Tolman explained that behavior of this kind is just the opposite of a passive response to a discrete stimulus, since it means that the subject is actively selecting the most relevant features of the situation and establishing relationships among them.

The third line of argument was based on an experiment by Hudson (unpublished thesis quoted by Tolman, 1948) involving the so-called "Search for the stimulus" and showing that avoidance conditioning cannot be learned if negative reinforcement occurs at the same time as the relevant stimulus disappears from sight. This result is consistent with the finding that rats undergoing normal conditioning, after receiving an electric shock seem to wander round the apparatus looking for its source. Tolman's interpretation of these data was on similar lines to that mentioned above: that animals engage in active searching and stimulus selection rather than being passive participants.

The "Hypothesis experiments" conducted by Krech (1932) provided the arguments of the fourth kind in favor of the idea of cognitive mapping. Krech demonstrated that in the course of complex discrimination learning, rats display different strategies successively until they find one that produces the optimal level of reinforcement. For example, during a lengthy training course, a rat in a maze begins by choosing only right-hand doors, then only left-hand ones, and so on. Tolman's explanation was that each strategy may correspond to the use of a tentative cognitive map.

Finally, "Spatial orientation experiments" provided the last type of argument and were aimed at demonstrating the existence of broad maps, as opposed to strip maps. In a narrow map or "strip map", two places, such as the starting point and the goal, are linked only by a single path. The usefulness of this map is therefore limited to a narrow strip of space, and it will be unsuitable when a wider area has to be crossed. If a change occurs in the initial situation (for instance, if the usual pathway is blocked), then it becomes necessary to use a broader map to solve the problem.

The data obtained by Tolman, Ritchie, and Kalish (1946) using the "sunburst maze" were taken by Tolman as evidence that rats are able to acquire more than a strip-map. In the first phase of the sunburst maze study, the animals learned to run from starting point A to the food-box G (Fig. 1.4). When they had become familiar with this route, test trials were introduced in which the apparatus was modified as shown in Fig. 1.5.

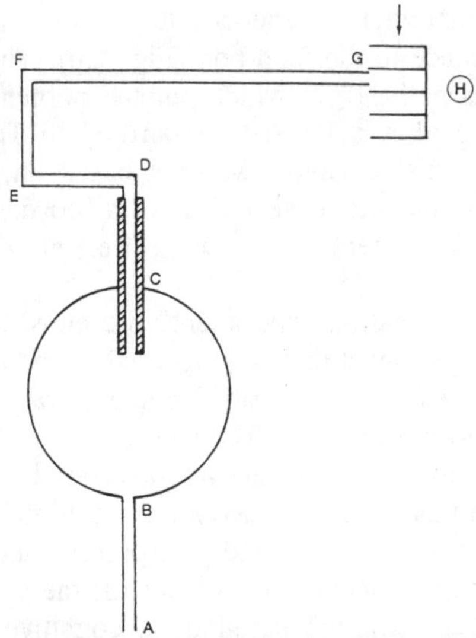

Fig. 1.4. Apparatus used in preliminary training (from Tolman, Ritchie, and Kalish, 1946).

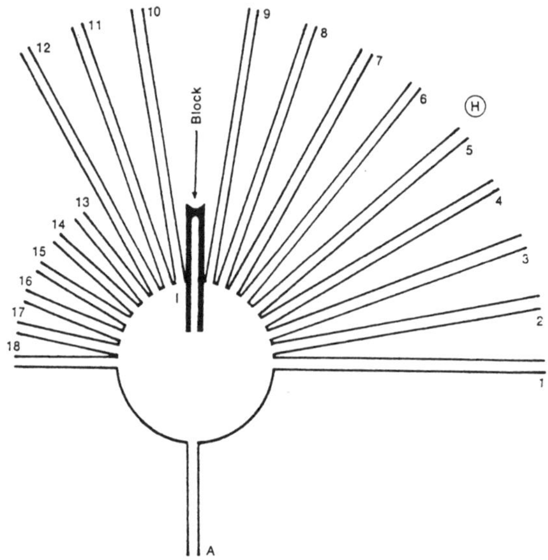

Fig. 1.5. Apparatus used in the test trial (from Tolman et al., 1946).

The initial pathway was blocked and numerous radiating paths were added. The rats showed a preponderant tendency to choose arm n° 6, which pointed toward the initial entrance to the food-box (Fig. 1.6). The other path which was frequently chosen was path N° 1, which pointed perpendicularly to the side of the room where the goal was located. According to Tolman, this experiment demonstrated that the rats' behavior was not based on the use of a strip-map limited to a single path but involved the use of a broader spatial representation; they knew in a more fundamental way *where* (i.e., at what place) the food was located in the room.

Another "Spatial orientation experiment" presented by Tolman was a study by Ritchie (1948), who showed that rats released from an unusual starting point in a maze were able to choose the path leading towards the side of the room where the goal was located (Fig. 1.7). This behavior cannot be explained in terms of S-R associations since the rat, when released from an unusual starting place, cannot have previously established an association between a stimulus (for instance the crosspoint of the maze) and a response (such as a right turn during the initial learning phase). According to Tolman, these data again demonstrate the existence of "place learning" based on a cognitive map, one purpose of which is to inform the animal about the location of the goal, regardless of whatever turns have to be made to reach it.

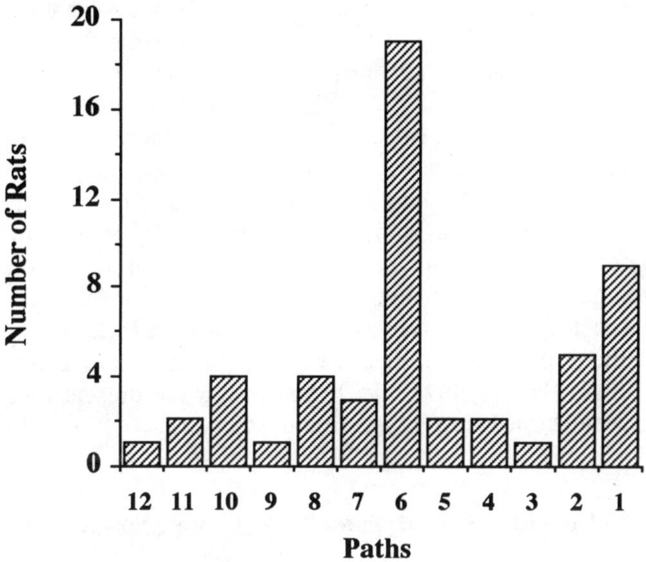

Fig. 1.6. Number of rats which chose each of the paths (from Tolman et al., 1946).

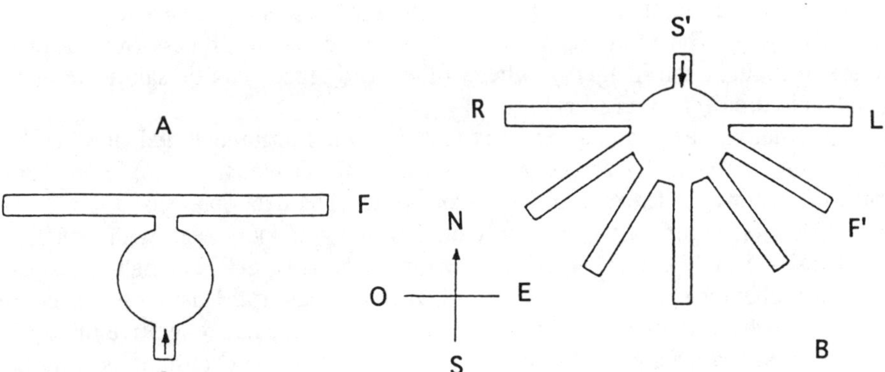

Fig. 1.7. Apparatus used by Ritchie (1948) in (A) preliminary training and (B) the test trial. For instance animals, starting from S, were trained to find food at F during preliminary training. For the test trial, the apparatus was rotated through 180° and a series of radiating paths was added. The goal F' was exactly as the same place as during preliminary training. The animals were released from S'. If during the test trial, the rat repeated a right turn as during preliminary training, it would have arrived at R.

On the basis of these arguments, Tolman described the main features of cognitive maps, which involve radically different mechanisms from those necessary for establishing S-R associations: first, maps result from the central processing of spatial information regardless of the motivational state of the organism and the presence/absence of reinforcement. Secondly, maps require that information be selected and organized in such a way that spatial relationships are established between various environmental elements. These relationships are far more important than the proprioceptive information of the kind involved in simple response learning. Thirdly, the maps need to be broad enough to be compatible with a relatively large degree of freedom.

Several important questions have still to be answered, however, as to whether these features provide an exhaustive account of cognitive maps, and how we should regard the cognitive map hypothesis at the present time in the light of what has been learned since it was first put forward.

2. How Should the Cognitive Map Hypothesis be Viewed Nowadays?

Although we have come a long way since the polemical debates which were waged among learning theorists in the 1950s, one central problem is still there: it concerns the fact that the term "cognitive map" was not really defined by Tolman who only alluded to the map's contents and properties. The expression sounds quite pleasing but is fundamentally antinomic and can easily lead to misunderstandings. The term "cognitive" refers to a dynamic process (mental processes of thought and imagery), whereas the word "map" can be said to refer to a static picture of the real world.

The main weakness, however, is that Tolman's arguments led only to the conclusion that spatial learning involves mechanisms which are *different* from those underlying the learning of S-R associations. The data obtained in the "Spatial orientation experiments" (the only ones which genuinely dealt with "spatial" problems) certainly cannot be accounted for in terms of S-R associations, since the rats' behavior did not depend on a rigid motor pattern. This does not, however, necessarily imply that there is only one alternative, i.e., cognitive maps, to the use of S-R associations. Moreover, a closer look at the paradigm used in the sunburst maze (Figs. 1.4 and 1.5) shows that a light cue was located just above the goal. As pointed out by O'Keefe and Nadel (1978) it may have been this light, rather than place discrimination, which led the rats to choose the most direct route to the food! By associating the light and the goal, the rats just needed to reduce the distance between themselves and the light in order to perform the task efficiently.

One of the advantages to be gained from using cognitive maps is that they enable an individual to reach a place from an unfamiliar starting point (see

Chapter 3), but it is unlikely that any mapping will be carried out if a conspicuous cue is directly associated with the goal. In this respect, the second experiment by Ritchie (1948) appears to be the most powerful argument supporting the cognitive map hypothesis (although it must be said that we have little information about the exact experimental layout!). The other data quoted by Tolman provide at best some indirect arguments. They certainly confirm that cognitive information processing takes place but only serve to determine *some* of the conditions which have to be fulfilled for the building up of cognitive maps to be possible.

Surprisingly, Tolman did not cite the experiment that he carried out with Honzik (1930), which might have supported his theory more convincingly. In an elevated maze, three routes (Fig. 1.8) led to the goal. When the direct route (1) was blocked at A, the rats retraced their steps and chose route 2. But when they discovered that it was blocked at B, they then chose route 3, which was the only one with which they could reach the food. These data, like those obtained in the "Spatial Orientation experiments", point to the crucial distinction that Tolman did not explicitly mention, namely the distinction between places and the possible paths which link them. Tolman and Honzik's data throw light on an important feature of the maps, namely the fact that the representation of spatial relationships must be *independent* of the pathways which are usually taken between them. The plasticity of cognitive maps arises from these characteristics but Tolman focused on the scope of the map. The fact that it is possible to use more than one pathway between locations means that the rat's representation is not limited to a single link between the starting point and the goal. Tolman did not refer to some clever previous experiments such as those conducted by Shepard (1931, 1933; see Introduction) which would have provided additional significant grounds for his hypothesis. Instead, he confined himself to arguments based on his own data and those of his collaborators.

Tolman's assertions are largely based on intuition, and in truth, hardly deserve the title "hypotheses". The terms in which his 1948 article was written bear this out: "My argument will be brief, cavalier and dogmatic", We believe ...", "We feel ...", "We assert ...". These were not the intuitions of a genius: some fairly sophisticated ideas about spatial representations had been put forward previously, many years earlier, based on a mixture of philosophy and common sense. But it must be said to Tolman's credit that he gave the topic of spatial representation in animals and humans its scientific status and respectability. In modern terms, one might say that Tolman put forward an alternative, in terms of "propositional memory", to the behaviorist explanation which focused on what we might now call "procedural memory" (Dickinson, 1980).

Looking back, it is difficult nowadays to evaluate Tolman's contribution because of the unavoidable hindsight provided by the data and theoretical developments that have accumulated from that time up to now. The main weakness of

Tolman's definition seems to be that it tells us very little about the map's properties. Maps are psychological constructs which can never be directly observed. Besides investigating their constituent elements, it is therefore necessary to adopt a constructive approach consisting of actively exploring some of their *functional* properties.

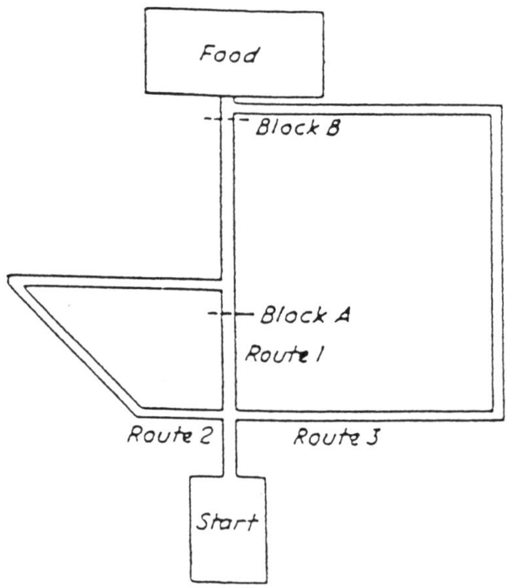

Fig. 1.8. Apparatus used by Tolman and Honzik (after Tolman and Honzik, 1930).

3. After Tolman: the Conciliation

After the publication of Tolman's paper, the cognitive map hypothesis was eclipsed up to the seventies and very few studies on this topic were published. Tolman's collaborators published a series of articles entitled "Studies in spatial learning" which was completed by Ritchie, Aeschliman, and Pierce (1950) and Ritchie, Hay, and Hare (1951). These studies included some statements which watered down the inordinately heavy-handed statement of the initial hypothesis. They also allowed the resarchers to escape from a conceptual bottleneck since they suggested that response and place learning need not be mutually exclusive (see Thinus-Blanc, 1984, 1987, 1988a, for further discussion). Based on Ritchie, Hay, and Hare's data, Restle (1957) suggested in a theoretical article that the probability of occurrence of either kind of learning might depend on the situation. In particular, the presence of a large number of relevant environmental cues might favor the setting up and use of cognitive maps, whereas response learning might be more likely to occur in a stark and featureless environment. Restle's hypothesis was confirmed experimentally by Blancheteau (1969), who conducted a series of experiments in elevated mazes (providing environmental cues) and enclosed mazes (the high walls of which prevented the rats from visually perceiving the surroundings). In addition, Blancheteau demonstrated that even in a diversified environment, the repetition of a pattern of movement initially relying on place learning induced response learning. This suggests that S-R associations may be made whenever they prove to be useful.

The theory recently developed by Toates (1994a, 1994b) about "what is cognitive and what is not cognitive" supports the idea that a hierarchy among mechanisms which control behavior does not imply that they are mutually exclusive. Rather, prevailing circumstances (as in Blancheteau's experiments, for instance) determine the implementation of such-or-such level of processing which may not necessarily be incompatible with other ones.

Our experience of everyday life provides many examples of the use of processes which are economical in terms of "cognitive load". For instance, the changes of direction occurring during the iterated journey from one's home to the office do not require recalling a map even if one has been constructed. Here the subject simply relies on a set of automatisms which result in "turning left at the traffic light", "turning right at the church", etc.

However, the fact that in some situations the same individual is able to use place or response learning, depending on the environmental layout, does not necessarily mean that this individual is capable of place learning in other more complex situations. It is therefore important to establish whether S-R associations (or other simple strategies), when they are observed, involve the highest level of the subject's capabilities. If not, they will serve an adaptive purpose since they constitute an economical means, in terms of "cognitive energy", of moving around efficiently. In the meanwhile, cognitive abilities are released for purposes other than the computation and planning of a trajectory. Anyone who has narrowly missed colliding with a car while driving and trying at the same time to find one's way will agree on this point. Likewise, animals under natural conditions need to give all their attention to detecting predators or other potential dangers. In addition, movements based on S-R associations are faster since the decisions are taken on the spur of the moment and involve neither computation nor planning ahead.

The use of stimulus-response associations amounts in this case to an adaptive procedure where one possible choice is made among a larger behavioral repertoire. As soon as this choice proves to be unsatisfactory, the subject will test another means of orienting.

Alternatively, S-R implementation may be the best available method for an individual coping with a spatial task. If it proves to be inappropriate, because of a change in the initial spatial configuration for example, then the behavior is discontinued and another strategy is used. The subject may be incapable, because of cognitive limitations, of carrying out higher level spatial processing. S-R associations can therefore either constitute 1) an adaptive strategy or 2) the ceiling level of ability. In the latter case, however, the possibility that higher cognitive levels may be involved can be ruled out.

Conversely, the fact that an animal succeeds a "spatial test" that has been specifically designed to prove that a cognitive map both exists and is used is usually interpreted in terms of its ability to construct appropriate representations, although it is often possible that high level performance may be based upon simpler strategies, such as perceptually guided behaviors involving the use of cues other than those intended by the experimenter (as in Tolman, Ritchie and Kalish's sunburst-maze), or S-R associations, or other still unknown means. This possibility is usually not willingly taken into account by most investigators. Likewise, when a subject fails in a spatial test, the tendency to oversimplify the explanatory framework by exclusively referring either to place or response learning has discouraged investigators from analyzing the *nature* of the errors committed by the subjects. Further analysis might however have thrown some light on whether some intermediate process was involved which was liable to account for the animals' low performance levels.

In conclusion, the conciliatory theoretical article by Restle (1957) and the experimental approach by Blancheteau (1969) – less well known since it was published in French – have appeased the factions in the Hull-Tolman controversy. It is granted by modern psychologists such as Toates (1994a), for instance, that various means of orienting, such as place and response learning, may coexist in the same individual, depending on the requirements of the task and the subject's cognitive abilities. Some serious data interpretation problems still remain to be solved, however. The problem is basically to disentangle "what one can do" from "what one would normally do" (Downs, 1981). In this connection, "what one can do" refers to the ability to construct and use a map. "What one would normally do", may involve implementing simpler strategies, such as making S-R associations, when these work efficiently. One may wonder whether the dualistic interpretative framework has not had a withering effect on the imagination of the investigators who have been content with this narrow dichotomy. Do there exist only two classes of mechanism liable to account for spatial orientation? Or is it reasonable to assume that some other modes of processing might be involved?

4. Toward a Multidisciplinary Approach to Animal Spatial Cognition

Up to the seventies, the interest shown in the cognitive map hypothesis was lulled for a while as if Restle's 1957 conciliatory article had definitely settled the debate. What was going on during that lull? Behaviorism predominated over psychology until the sixties or thereabouts, after which cognitive psychology came gradually to the fore. Many studies on human subjects, dealing with attention, perception, memory and language, contributed to forming the widely-held idea that the subject actively processes and stores information and is able to construct representations. Piaget's theory about the development of intelligence and spatial knowledge also represented an invaluable contribution to the flowering of cognitive psychology, although his books were translated into English long after being originally published in French.

All these studies were carried out on human subjects, but they delineated a sufficiently general theoretical background to be applicable to animals as well. For instance, in their studies about environmental cognition, Lynch (1960) and Appleyard (1969, 1970) do mention the notion of maps. It is likely that the development of human cognitive psychology may have induced animal specialists to take a new look at the cognitive map notion.

Around the seventies, especially during the period between 1976 and 1981, animals' spatial cognition attracted renewed interest which is still continuing. Several important events took place at that time. First, in 1978, O'Keefe and Nadel published their famous book "The Hippocampus as a Cognitive Map" (see below), which was a significant milestone in the multidisciplinary approach to spatial representation in animals. Secondly, two popular behavioral tests came out during this period. These two tests were extensively used by their authors, and subsequently by many other research workers, to evaluate the contribution of various brain regions, especially the forebrain hippocampus, to spatial processing and the effects of pharmacological treatments. The first of these tests was that based on the radial arm maze which was designed by Olton and Samuelson (1976). What made this experimental situation so original was that it included several baited, perceptually similar goals (one at the end of each arm), which had to be visited in turn without returning to any of the previously visited (i.e., emptied) places. The animal was free to adopt any pattern of visits to the goals. By analyzing the sequence of choices, some insights were obtained about the spontaneous organization of the behavior. In addition, since the correctness of any one choice in the sequence of visits depended upon the previous ones (in order to avoid visiting emptied arms), this situation made it possible to adopt a more comprehensive approach than previously to the analysis of choice behavior. It encompassed both the short-term storage of the action and the long-term memory of the spatial relationships between each arm and the environmental cues used to identify the arms' respective locations.

A second experimental set-up that also led to considerable advances in this field was Morris' Water-Maze (Morris, 1981). This experimental device was specifically designed to analyse the properties of cognitive maps. In addition, testing animals in water made it possible to overcome some of the drawbacks of testing in "dry" mazes, such as the possibility that the animals may have made use of olfactory scent trails and the necessity to deprive the animals of food to an optimal level to control their motivation. These problems were eliminated in the water maze, since the animals could swim and were strongly motivated to reach the firm platform serving as the goal.

Together with other less widely publicized but sometimes equally fruitful studies which were published during the seventies, several authors using these popular experimental methods (which are described in detail in the following chapters) contributed greatly to the analysis of animals' cognitive maps. Moreover, they significantly enriched the conceptual and theoretical background to the study of animal spatial cognition. Perhaps the most noteworthy feature of Olton, Morris, and O'Keefe and Nadel's approach, however, is that it integrated into a unified framework both the behavioral and brain mechanisms involved in spatial representation. For this reason, these authors can be said to have launched a remarkable and representative example of the *cognitive neuroscience* approach even though this discipline had not yet been christened as such at that time. On the one hand, the creation of new experimental designs made it possible to further the understanding of spatial processing at the psychological level and provided subtle and refined tools for investigating the brain's mechanisms. On the other hand, advances in the understanding of how the brain functions contributed to developing and refining hypotheses at the psychological level. The whole range of means available for analyzing and differentiating between the various constient elements of highly complex spatial behaviors was therefore enlarged.

Joint approaches to the neuronal and behavioral mechanisms underlying spatial orientation are not, of course, a recent development. As early as 1920, Lashley started testing animals with various brain lesions in spatial tasks. However, exactly as happened in the field of behavioral psychology, this approach was not systematically applied to closely determining the neural bases of spatial cognition. From this point of view, "The hippocampus as a Cognitive Map" (O'Keefe and Nadel, 1978) was a significant landmark in the multilevel and pluridisciplinary approach to spatial representation. This book has had a considerable impact and is still bedside reading for many specialists in the field. Among the many reasons for its success, the first was that the authors filled a gap, namely the absent broad theoretical framework. They went beyond Restle and Blancheteau's conciliatory conclusions: they integrated several classes of psychological mechanism and data from various domains into a coherent theoretical framework. In particular, they proposed a distinction between two classes of orienting system: the "Taxon" and the "Locale" systems. Let us first consider the behavioral side of the theory.

The Taxon systems involve the use of "routes" which "... direct the subject's attention to particular objects or specify turns within the egocentric space." (p. 83). These "routes" can be viewed as lists of successive stimulus-response-stimulus (S-R-S) instructions. Within each S-R-S instruction, the most prominent element can be either the stimulus itself or the response requirement. In the former case we have "guidances", and in the latter, "orientation" and "direction". O'Keefe and Nadel defined guidances as follows:

"*A guidance directs attention to a particular landmark or object and requires that the rambler approach it or maintain a certain egocentric relationship to it, regardless of the behaviour involved.*" (p. 82).

The particular landmark can be either the goal itself or a subgoal. The behavior consists of reducing the distance between the subject and an attractive perceived goal, or conversely, moving away from an undesirable element that is to be avoided. In some cases, environmental stimuli can be associated with a goal which is not perceived (as in the sunburst maze, for example).

In the cases where the response requirements predominate over the cue aspect, routes are followed on the basis of "orientations" or "directions". The responses consist of making rotations within the egocentric space so as to align the body axis in relation to some other axis. The simplest case consists of making a rotation relative to an initial orientation and in continuing straight ahead. A more complex route is characterized by a sequence of successive re-orientations of the body axis. Since it may be difficult, however, to make turns and to walk straight on the sole basis of proprioceptive cues, O'Keefe and Nadel proposed that a distant landmark may be used as a directional cue. Although this directional cue may look very similar to a guidance, there are differences which become obvious when the subject gets lost or has to make a detour, for instance. When a guidance is implemented, the only means of finding one's way is to wander at random until the guide is encountered. The problem is even more difficult, if not insoluble, in the case of orientation. Directional cues are useless here because they provide orientation in relation to the body axis, and that is the very information which has been lost.

As a matter of fact, the most distinctive feature of routes is their serial structure. Each segment between two subgoals or between two re-orientations has its own functional role because it comes *after* the previous ones and *before* the next ones. This explains the lack of plasticity of routes, as O'Keefe and Nadel (1978) pointed out. If one subgoal has disappeared or has not been noticed by the subject, or if the subject's path has deviated slightly, then the whole sequence is disorganized and interrupted, since one link in the chain is either missing or inappropriate. In addition, the context is not taken into account while following a route. Therefore, no distinction will be made between two identical guides belonging to different contexts, and this may lead to inappropriate behavior. Routes do have one advantage however which is the rapidity of their execution since they require very little information processing.

The Locale system relies on the construction and use of maps. O'Keefe and Nadel define a map as "... the representation of a set of connected places which are systematically related to each other by a group of spatial transformation rules" (p. 86). Angle and distance relationships are assumed to be approximately encoded in this type of representation, although they may suffer from some distortions.

According to the authors, one of the most important properties of maps is that places are not defined in terms of the objects which characterize them. They admit that it is convenient to locate a place by referring to the object occupying it, but "... this is only a convenience" (p. 86). Maps refer to an absolute space, as defined by the nativist philosopher Immanuel Kant (1787), which may be empty and exists whatever elements it may contain. Accordingly, this form of representation is completely independent of the subject's position at a given moment. Unlike the Taxon systems, spatial relationships are not organized with respect to the subject's position (egocentered) but rather by referring them to a fixed and absolute (allocentered) frame of reference.

Maps are endowed with other more functional properties such as flexibility. Unlike routes, which depend upon the performance of specific types of behavior, maps can be used in many different ways to go from one place to any other, since the places are not linked together by particular movements but rather by various truly *spatial* relationships. Consequently, the behavior can be adapted to constraints such as the presence of an obstacle necessitating a detour. They are also resistant to some loss of information because of their high information content and redundancy. For this reason, they take longer to use, since a large amount of information has to be processed and several options may be available. In addition, since the representation of a place on a map is encoded in terms of its spatial relationships with other elements, no confusion is possible between similar elements belonging to different situations. The information is "context-specific".

A further crucial element on the psychological side of O'Keefe and Nadel's theory was the spatial function of exploratory activity which they conceived of as "... behaviour designed initially to build and subsequently to update cognitive maps" (p. 94). Since there is no map of an unfamiliar environment, curiosity triggers a set of investigatory behaviors, the main function of which is to set up spatial relationships between places and to build up a map of that specific situation. Likewise, if a novel element is introduced into a familiar situation, the detection of the incongruence between the initial map and the actual situation activates the cognitive and motor systems involved in exploration in order to update the map by incorporating the change.

O'Keefe and Nadel's definition of maps is quite in keeping with the conceptual framework delineated by Tolman thirty years earlier but they go well beyond Tolman's rather crude theorizing. Although their definitions were sometimes a bit blurred (for instance, what did they mean by the "group of spatial transformation rules" which link places together?), they provided some explicit guidelines for addressing the nature and properties of maps. In addition, the spatial function of exploration, which in 1978 was highly speculative, has since been experimentally confirmed.

The most original feature of the theory concerns the distinction between guidance and orientation. Instead of taking S-R associations to form a monolithic category of response learning versus place learning, O'Keefe and Nadel make the distinction between those behaviors relying predominantly on particular landmarks (guidances) and those where the response requirements are more significant (orientations). Although the two types of behavior are both organized in an egocentered reference frame (in this respect they are the opposite of maps), this distinction can be said to have paved the way for new hypotheses about intermediate means of orienting.

In addition to the psychological bases of oriented displacements outlined above, the other noteworthy aspect of O'Keefe and Nadel's theory is that they clearly stated which brain structure they took to be the substrate of the maps, namely the hippocampus. To support this hypothesis, they provided a detailed account of this anatomical structure and its physiological functioning and reviewed the behavioral effects of hippocampal lesions. Of course, they also extensively developed another compelling argument for the idea that the hippocampus serves a spatial function, i.e., the fact that it contains "place-cells", the properties of which were described first by O'Keefe and Dostrovsky in 1971. These arguments will be discussed in detail in Chapter 6. Briefly, O'Keefe and Nadel argue that map construction consists of three stages, each of which involves one of the three main regions of the hippocampus (the fascia dentata and fields CA3 and CA1). In the first stage (mediated by the fascia dentata), the various environmental inputs provided by the Taxon systems, which correspond to what the animal has successively experienced, are transformed into a unitary and global representation. During the second stage (which takes place in CA3), places in the environment and the angular and distance relationships between those places start to be represented. At the final stage (which occurs in CA1), the map is completed, and among other things, any new elements are detected ("mismatch system"). The slow wave rhythmic pattern found to occur in the hippocampal EEG, or theta rhythm, plays a crucial role in this scenario by imposing "... a synchronous bursting pattern ..." (p. 225) on neuronal subsets. It is said to "... shift the focus of excitation from one set of place representations to another within the map ... on the basis of the animal's actual or intended movements" (p. 224). This is the dynamic aspect of spatial orientation which had been neglected by Tolman and other investigators. As a matter of fact, knowing the spatial relationships between a starting place and a goal would be of no use if the animal were unable to keep track of its position while moving toward the goal. Some kind of updating of the animal's own position at any moment, in relation to the environmental layout, appears to be crucial for a trajectory which was initiated at the starting place to be successfully pursued up to the goal.

In addition to this highly simplified account of O'Keefe and Nadel's theory about the spatial role of the hippocampus, a few other features are worth pointing out which are at the core of current debates on spatial memory and will be referred to later in this book. First, these authors did not explicitly (or implicitly) consider the possibility that other brain structures might participate with the hippocampus in the construction and storage of cognitive maps, and secondly, they assumed that the hippocampus is the site where the long-term storage of the maps occurs. On these two points, the theory is at variance with recent hypotheses about distributed memory and the transience of the hippocampal contribution. The third important feature of O'Keefe and Nadel's hypothesis is the role that they attribute to the hippocampal theta rythm, which is more in line with present-day thinking. The hypothesis examined in Chapter 9, that the theta rhythm may impose temporal coordination on assemblies of neurons, is very germane to recent developments and data on the temporal coding of information in the brain.

We have paid special attention here to O'Keefe and Nadel's theory because it involved taking a great deal of data based on both brain and behavioral studies and integrating it into a single coherent conceptual framework. Other important studies on the spatial role of the hippocampus were carried out by many authors during the same period, however. This mainstream is still very much alive. In contrast with the huge corpus of studies which bears upon the spatial specialization of the hippocampus, only a few have dealt with the role of other brain structures in the processing of specifically spatial information. Although it is undeniable that the hippocampus plays a key role in spatial mapping, it is not really far-fetched to imagine that other brain regions might possibly be involved in the production of the complex behaviors which rely on map implementation. It is striking that in human patients, spatial memory disorders are usually associated with posterior parietal damage, whereas in rats disorders of this kind are associated with hippocampal dysfunction. It is quite possible that the hippocampus may have evolved from playing a specifically spatial role in rats to a more general memorizing role in humans. Moreover, the size of the rat's neocortical areas is embryonic in comparison to that of humans. Rats do, however, possess both a hippocampus and a parietal cortex and although the latter may be functionally less highly evolved in rodents than the relatively well developed archipallium, the spatial role of both this and other neocortical areas should not be neglected.

A broader view can be taken not only of the brain structures but also of the behavioral processes. Constructing and using a map are only two phases in a process which originates at a subject's very first confrontation with an unfamiliar environment, and may end up with a successfully oriented trajectory. As early as the first stage, sensory systems, especially the visual system, perform a type of spatial processing which is egocentrically organized since it depends on the subject's position in the environment. Likewise, an appropriately oriented trajectory requires the information charted on the representation to be *translated* into a sequence of actions which are also egocentrically organized. A perfectly accurate map would not be of any use if this kind of transformation could not be made. During behaviors which involve the construction and use of spatial representations, there must occur continuous back-and-forth exchanges between the processes directly involved in controlling actions within the physical environment and those which deal with the more abstract representation. It is necessary to take into account all these components if one wants to reach a better understanding of spatial processes in general, and cognitive maps in particular.

Adopting this approach to the problem has made it clear that several brain regions presumably come into play and cooperate at each of the several stages which lead to an oriented behavior. It is also obvious, however, that a careful behavioral analysis is required in order to tease out the respective contributions of all the brain regions involved. An analysis of this kind then becomes an actual tool for "dissecting out" and isolating the various components of an extremely complex process.

To summarize, the joint approach to the brain-related and behavioral bases of spatial cognition, as clearly stated by O'Keefe and Nadel, gives rise to a question which will be extensively discussed later on in this book: is the hippocampus the sole brain structure involved in spatial maps or are there other regions which may be directly committed to this level of processing?

5. Organization and Content of the Book

The aim of this book is to examine the questions which have been raised in this Introduction and to attempt to provide some ideas as to how they might be answered. I will start (Chapter 2) by dealing with the important relationship between spatial processing and exploratory activity. Several studies will be discussed, which have yielded data supporting the idea that the spatial features of the environment are actually organized and encoded during exploration. This approach, which is often neglected, may throw some light on the actual nature of spatial representations. The next chapter (Chapter 3) takes a critical look at the most classical experimental situations which have been designed so far to investigate the nature of the spatial representations that animals construct and use.

The question here is whether the various situations which have been extensively used, purportedly to determine the existence and properties of cognitive maps, actually do so.

Chapter 4 deals with a fundamental property of maps, namely their plasticity and their usefulness from the adaptive point of view when some unexpected change occurs in a familiar spatial layout. The ability to make shortcuts and detours can be taken to reflect the plasticity of maps and the existence of a form of encoding which is independent of whatever movements have taken place within a given environment. In Chapter 5, the various models which have been proposed to account for spatial processing at the behavioral and psychological levels are discussed. Data from lesion and electrophysiological studies which shed light on the spatial role of the hippocampus are dealt with in Chapter 6. The possible involvement of other brain structures, such as the posterior parietal, occipital and prefrontal cortices, in specific aspects of spatial processing is discussed (Chapters 7 and 8) in the light of the few available data (few in comparison with those available on the role of the hippocampus). In Chapter 9, the relevance of various models to specific spatial memory and mapping processes at the brain level is discussed. Lastly, in Chapter 10, some suggestions for future research in the domain are provided on the basis of the behavioral and brain data and the theoretical developments which have been reviewed in the foregoing chapters.

Although ethological studies have been carried out on numerous kinds of spatial orientation in a variety of animals, the data published in this field have not been included here except when they are particularly relevant to the questions under debate: the ethological approach is classically concerned with species-specific behavior, whereas this book is aimed at deducing some of the principles of spatial representation that can be generalized regardless of species. This does not mean that I consider the data obtained on rodents and other non-primate animals to be directly applicable to human spatial cognition. Some features of spatial processing do nevertheless seem to be common to both animals and humans. It would be a mistake, for example, to imagine that human spatial representations are always encoded in terms of language and high level symbolic processes. The path-integration processes, which have been found to occur both in various animal species and in humans, provide a good example of the possible similarities in the way spatial processing is carried out by both human and non-human species.

CHAPTER 2

EXPLORATION AND SPATIAL KNOWLEDGE

"Few items or places are completely novel; novelty typically consists in new configurations of familiar elements ... the novelty of the wife in the best friend's bed lies neither in the wife, nor the friend, nor the bed, but in the unfamiliar conjunction of the three." (O'Keefe and Nadel, 1978, p. 241)

All mammals react in some way or other to novelty. Exploratory behaviors are particularly obvious in rodents. A rat or a hamster placed in a new environment displays feverish investigatory activity, which finds expression in multiple contacts with objects in the environment, sniffing, stopping, rearing up on the hind paws, and so on. In human subjects, reactions to novelty, although different, are also observable. Young children displaying an irrepressible need to seize and touch any unfamiliar object within their reach, people entering a restaurant and looking around with discretion, provide convincing illustrations. One feature of this activity is its decrease over time, a phenomenon called habituation. Depending on the complexity of the situation, its degree of novelty, and the species concerned, the process of habituation varies.

Berlyne's development of ideas about curiosity and reactions to novelty (1960, 1966) has contributed widely to the belief that these behaviors are part of the general process of knowledge. A less widespread, though non-revolutionary, hypothesis is that exploration has a role to play in spatial knowledge. Available data provide compelling evidence of a spatial function for exploration. This chapter aims to examine this function from empirical and theoretical viewpoints.

1. Early Investigations

The spontaneous alternation task has been traditionally considered to represent a test of reaction to novelty. During consecutive trials, a rat is placed in the start alley of a two-choice T- or Y-shaped apparatus and is allowed to explore freely. Usually the animal on the second trial chooses the arm which was not visited on the first trial. This phenomenon is considered to be exploratory activity since it reflects the animal's tendency to investigate an unknown part of a new environment. This behavior shows that familiar locations are correctly identified and differentiated from unfamiliar ones. The tendency to alternate increases if the subject is kept for a while in the goal-box chosen in the first trial (Glanzer, 1953). In addition, this phenomenon does not rely on the alternation of different motor responses (e.g., to turn left then turn right) but on visiting distinct places, since when two paths lead to the same location, the alternation rate decreases (Sutherland, 1957).

However, the most popular situation in which to study exploratory activity is the open-field apparatus, designed by Hall (1934) to evaluate the effects of an unfamiliar environment on emotional behavior in the rat. In spite of its justifiable use in certain fields of research, the original version of the open-field test (consisting of an empty field, divided into segments by lines on the floor) does not provide much information on spatial knowledge. By merely recording the number of squares entered in a unit of time, rearings on the hind paws and defecation, highly instructive features of behavior go unrecorded.

A more fruitful method than the mere measurement of spontaneous exploratory behaviors in an empty arena is to record animals' responses to the placement of unfamiliar objects in the experimental field, or to the introduction of new objects into their living cage. Reactions to new objects were first studied by Berlyne in a pioneer experiment (1950): one of three familiar objects contained in a cage was replaced by a new object. In this situation, rats were seen to display investigative behaviors that were *selectively* directed toward the unfamiliar object. These reactions appear to be modulated by various factors such as the level of familiarity of the context and the conditions (enriched or impoverished) in which the animals were reared (Cowan, 1976).

Berlyne's key idea in his 1960 paper was that previously stored information regarding some characteristics of an object or of a situation serves as a reference for novelty detection. Novelty per se does not exist, but only through reference to familiarity, and vice-versa. The relativity of these two opposite features which characterize all the objects, events and thoughts that constitute any individual's life experience, has deep methodological and theoretical implications: methodological, because the dishabituation paradigm is now extensively used in many studies of shape recognition and concept formation in young children and animals; and theoretical because Berlyne pioneered the view that exploratory activity does much more than just reducing a primary motivation since it involves actual cognitive processing. This view is shared by others such as Renner (1988), Thorpe (1963), and Toates (1994b).

As explained in Chapter 1, O'Keefe and Nadel (1978) go even further in the role they attribute to exploratory activity since they assert that one of its functions would be to update cognitive maps. The less familiar a situation, the higher the degree of exploration. When confronted with a partly or totally novel situation, there is a systematic screening of stored memories of familiar environments. The detection of a discrepancy or mismatch between what is known or expected and what is currently perceived triggers exploratory activity, which is aimed either at constructing a map of a totally new situation or at updating an already internalized model to incorporate the present modifications. Then, habituation corresponds to the increasing degree of matching between the two situations (the stored and the currently perceived) that are being compared. Sokolov (1960, 1963a and b) had developed a similar analysis of the orienting response, this being one of the constituent elements of exploratory activity (the orienting response consists, inter alia, of a head movement in reaction to an unexpected stimulus or event appearing in the peripheral visual field). Gray (1982) and O'Keefe and Nadel (1978) have put forward the hypothesis that the hippocampus performs this comparison function. In fact, the involvement of the hippocampus in both exploration and in spatial memory (see Chapter 6) is a weighty argument for thcir having a rcciprocal rclationship.

2. The Dishabituation Paradigm

The methodological and theoretical interest of exploratory activity has not escaped the notice of specialists in animal psychology who have seen in the dishabituation paradigm an invaluable means of investigating a wide range of cognitive abilities in animals. Some of these studies deal specifically with spatial knowledge. Whatever the procedural details and species under study, the principle of these experiments is the following: subjects are repeatedly exposed to a given spatial configuration (usually one or several objects contained in an open field). Then after habituation of exploration has occurred, some spatial relationships between the elements of the situation are modified (for instance by displacing one object). What is new is not the individual characteristics of the objects since these have been previously investigated, but rather their spatial arrangement. If the animals react to the change by displaying a renewal of exploratory activity, then it can be assumed that the detection of spatial novelty relies on a comparison between the currently perceived arrangement and a representation or stored "internal model" of the initial situation.[a]

Several species have been tested using this procedure: hamsters (Poucet, Chapuis, Durup, and Thinus-Blanc, 1986; Thinus-Blanc, Bouzouba, Chaix, Chapuis, Durup, and Poucet, 1987), mice (Ammassari-Teule, Tozzi, Rossi-Arnaud, Save, and Thinus-Blanc, 1995; Misslin, 1983; Roullet and Lassalle, 1990), Mongolian gerbils (Cheal, 1978; Wilz and Bolton, 1971), rats (Poucet, 1989) and monkeys (Menzel and Menzel, 1979; Joubert and Vauclair, 1986).

---

[a] The reaction-to-change paradigm can also be used as a visual discrimination test by modifying after habituation some visual but spatial features of the situation (e.g., Thinus-Blanc and Foreman, 1993).

All display renewed exploratory reactions to a spatial change, which provides evidence that animals of all these species had formed an internal model of the initial situation.[b]

It is possible to go further in the analysis of the internal models or spatial representations using the dishabituation paradigm. In fact, it allows us to glimpse some of the spatial features of a situation that are spontaneously encoded during exploration, since the various types of rearrangement differ in the reactions they evoke.

In a series of experiments (Poucet et al., 1986; Thinus-Blanc et al., 1987), hamsters were given three 15-minute sessions of exploration several hours apart. The apparatus was a circular open field containing four different objects, located within a curtained environment that was visually homogeneous except for a striped pattern. This set-up was designed to induce the animals to focus their attention on the objects contained in the experimental field by preventing them from being distracted by irrelevant distal cues. The number and duration of their contacts with the objects were recorded. After habituation had occurred from the first to the second session, a spatial change was made in the initial situation by displacing one (or several) of the objects. For the control groups, the arrangement remained unchanged during the third session. Three types of behavior were observed to occur: (1) re-exploration was selectively directed toward the object that had been displaced. This occurred when one object was set apart from the others (Fig. 2.1 left). The same pattern of results was obtained when the objects were identical. (2) Re-exploration was directed toward all of the objects whether they had been displaced or not. Such non-selective behavior was observed when one object was moved closer to the others (Fig. 2.1 right) even with initially large distances between the objects. Hamsters reacted as if the situation was completely new. (3) No re-exploration was observed, as when the four objects were displaced equally so as to form a square larger than the initial one. The reverse modification, whereby the size of the initial square was reduced, did not elicit a greater reaction. It is unlikely that such a massive change was not detected by the animals which had been able to detect far more subtle modifications in other tests. Therefore, a process of generalization may have occurred, preventing the animals from reacting to a change which modified neither the shape of the arrangement nor the topological relationships between the objects.

Several specific conclusions can be drawn from these experiments: First, hamsters appear spontaneously to encode geometrical relationships since they always react, selectively or not, to a change bearing on this parameter. This

---

[b] The habituation rate may vary according to the species under study (Poucet, Durup, and Thinus-Blanc, 1988; see also Thinus-Blanc, Save, and Poucet, 1996a, for a discussion about methodological issues).

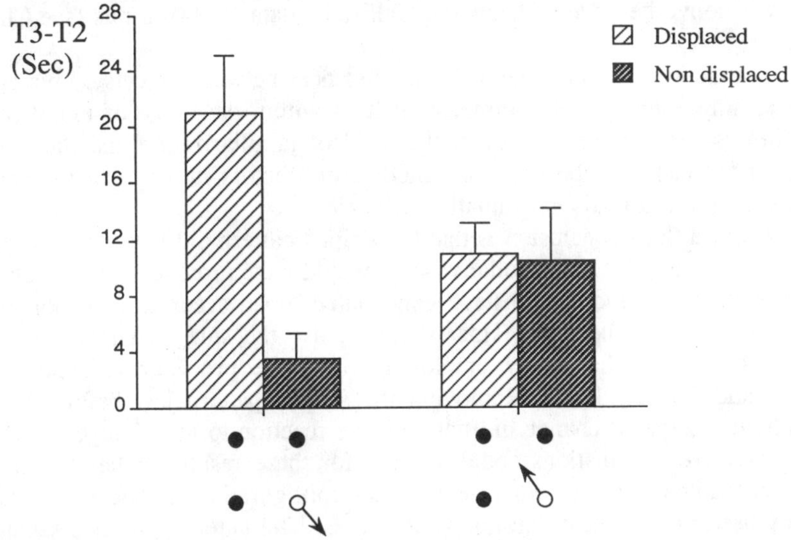

Fig. 2.1. Differences between sessions 3 and 2 in mean time (in seconds) spent investigating either displaced or non displaced objects (standard errors are represented by vertical bars). (Adapted from Thinus-Blanc et al., 1987.)

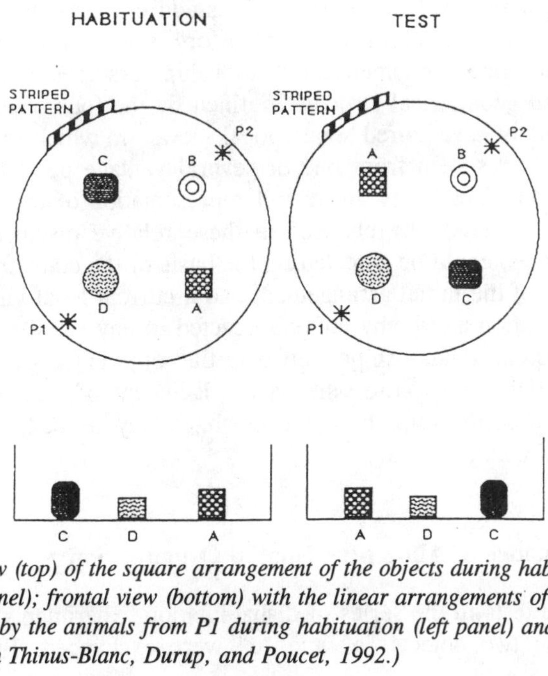

Fig. 2.2. Top view (top) of the square arrangement of the objects during habituation (left panel) and test (right panel); frontal view (bottom) with the linear arrangements of the objects as they can be perceived by the animals from P1 during habituation (left panel) and test (right panel). (Reproduced from Thinus-Blanc, Durup, and Poucet, 1992.)

finding is convergent with Cheng and Gallistel's data (1984) in rats (see Chapter 5).

A second conclusion is that absolute distances between objects do not appear to be an important spatial parameter, at least within the range of modifications used in this experiment. However, the level of familiarization and the requirements of the task may be important modulating factors in terms of the abilities displayed by the animals in evaluating distances.

Finally, a third conclusion is that the striped pattern did not appear to play a role in the novelty detection; otherwise, it would have provided the animals with a stable reference and would have contributed to an accurate detection of the change in most of the tests. This failure to use the sole visual cue has been confirmed in a further study employing the same procedure (Thinus-Blanc, Durup, and Poucet, 1992). Removing the striped pattern just before the test-session with a spatial change did not alter the reaction to this change. However, two alternative explanations could account for these results: either the animals relied exclusively on the initial geometrical configuration to detect the change, or they used uncontrolled extra-apparatus cues. The latter hypothesis seems unlikely given the short period of time the subjects were exposed to the situation, and the lack of any behavioral constraints. In addition, using such cues would have made the selective detection of the displaced objects very difficult because of the imprecision of this distal information.

At a more general level, these conclusions address theoretical issues related to the processing of spatial information. More specifically, concluding that animals appear to encode geometrical relationships assumes that they have extracted the overall geometrical structure defined by the four objects. Equally, it could mean that they have stored some "local views" in which relative distances between the objects as seen from one or several vantage point(s) are precisely encoded. This would not imply an overall representation of the object arrangement. Modifications involving changes in these relative distances (and hence, geometrical changes) could be detected on the basis of the comparison between a stored local view of the initial arrangement and a current local view. This might provide an explanation as to why animals reacted to any small change affecting the relative metrical relationships between the objects (Figs. 2.1 and 2.2) whereas they did not re-explore when a displacement of the four objects left their relative relationships (and therefore the shape they defined) unaffected. The next section deals with this issue.

3. The Predominance of Allocentric Spatial Organization

In one of the tests in the series of dishabituation experiments (Thinus-Blanc et al., 1987), the two objects "a" and "c" were exchanged (Fig. 2.2, upper panel). Unexpectedly, the hamsters reacted accurately to this change, selectively re-exploring the exchanged objects though the overall square shape defined by the object-set was unaffected.

In these situations, the four objects contained in the field define a square for the experimenter who has a bird's-eye view of the situation. However, the animal in the field has frontal views of this configuration which visually varies as a function of the subject's current location. Consequently, a more realistic way of representing the arrangement in which hamsters confronted the object set is displayed in Fig. 2.2 (lower panel) which shows the linear arrangement of the objects as seen from location P1. Location P1 was the place where the animals were consistently introduced into the apparatus at the start of each session and where they were often observed to stay for a while, displaying visual scanning, before moving around the arena. In the lower right panel of Fig. 2.2, it appears that an animal using location P1 as a particular viewpoint would be greatly helped in detecting the change and locating it with accuracy, by simply comparing the linear order rather than the overall geometric pattern of the object arrangement before (Fig. 2.2, left) and after the change (Fig. 2.2, right): object "a" which was initially on the right of the animal is now on its left during the test with the reverse being true for object "c". Therefore, it could be hypothesized that, instead of using an overall "bird's-eye-view" map of the spatial object configuration, hamsters may have constructed and used a specific representation of the spatial array as seen from a particular location such as point P1.

This possibility appears to raise an important issue inasmuch as it points to the type of frame of reference (i.e., egocentric vs. allocentric) used by the animals to locate the change. As a matter of fact, the implications of this issue are of concern vis-a-vis the conclusions which can be drawn from such experiments, in relation to the processes involved in gathering and reorganizing information about a spatial representation. If the detection of a spatial change following habituation is based upon a comparison with a stored spatial representation that depends on a particular place and viewpoint, then it should not be concluded from all of the studies cited above that an overall representation or map has been constructed during exploration, since, by its very nature, a map holds information regardless of the subject's specific position (O'Keefe and Nadel, 1978).

This issue has been addressed in a study in hamsters by Thinus-Blanc et al. (1992). After habituation, the two objects "a" and "c" were exchanged as shown in Fig. 2.2 (right) and the animal was placed in the field at P2. In spite of a clear tendency to react to the change, the dishabituation score did not reach statistical significance (Fig. 2.3, upper panel). However, close examination of the results revealed that 6 out of the 10 hamsters did selectively react to the change. Two other subjects re-explored both displaced and non-displaced objects whereas the two remaining hamsters failed to display any reaction. This diversity of strategies stands in contrast with the consistency of the results obtained when the

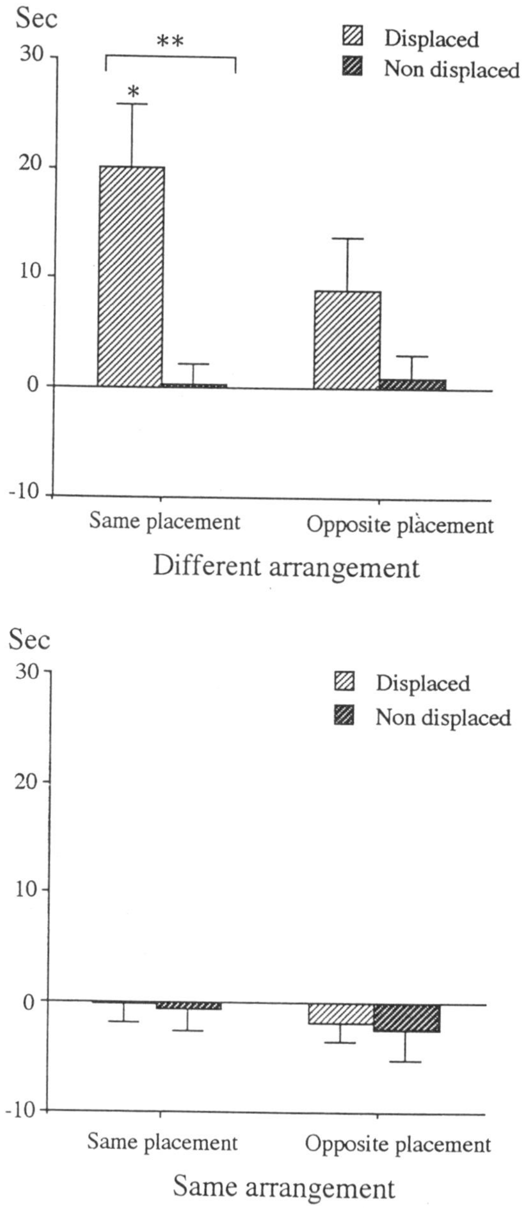

Fig. 2.3. Differences of exploratory activity (in seconds) between the test (two objects interchanged) and the last session of habituation in the group with the same initial placement in the field for both sessions and in the group with an opposite placement. (Adapted from Thinus-Blanc et al., 1992.)

animals where introduced in the field at P1 for both the habituation and test sessions. In addition, no reaction to a different visual alignment was observed in a group placed at P2 after habituation but without any exchange of the objects (Fig. 2.3, lower panel). Had the animals relied exclusively on a local view, changing the location of their introduction into the field should have elicited reactions to the perceived novelty. Therefore it may be that the heterogeneous results from these animals are related to the use of different strategies. The examination of individual results showed that a few animals seemed to have relied upon a particular local view while the majority of them had set up more global representations, independent of their current position.

These data suggest that the building up of spatial representations may go through several stages. Reaching the final level of exocentric cognitive maps could depend on the time allowed to acquire information, on the subject's cognitive abilities and finally on the requirements of the task. Indeed, further experiments using more complex spatial arrangements, with visually occluded objects, for instance, and different durations of exploration are required in order to determine the various phases, if any, of map construction.

On another hand, there is a body of data which unambiguously demonstrates rodents' ability to react to a change using allocentered strategies. For example, Tomlinson and Johnston (1991) conducted an experiment with hamsters using two visually identical objects differing only on the basis of their olfactory characteristics. After habituation, the two objects were both associated with the olfactory cue previously used for one of them. Thus the novelty can be described as "a familiar odor at an unusual place". When introduced in the field at a point 180° from the point of entry for the habituation phase, hamsters reinvestigated the object whose odor had been changed.

Thinus-Blanc and Ingle (1985) demonstrated that gerbils are also able to use allocentered strategies. After habituation had occurred with an object located in a field containing a sole striped pattern, another object similar to the first (including the olfactory scents deposited on it) was located symmetrically to the first with regard to the striped pattern (see Fig. 2.4). While gerbils had access to the apparatus through door D1 during the exploratory phases, they entered the field through D2 for the test. In spite of being provided with a different vantage point at the very beginning of the test session, they accurately re-explored the object which was new on the basis of its location only (Fig. 2.5).

These data are convergent with those obtained in rats by Xavier, Saito, and Stein (1991) and Poucet and Buhot (1989). In both experiments, the animals displayed reactions to a change even if they approached the locus of this change from opposite directions.

Other situations in the same vein have been devised, relying on habituation and dishabituation following a spatial change. The distinctive feature of these experiments is that whereas the initial environment that the animals explore is

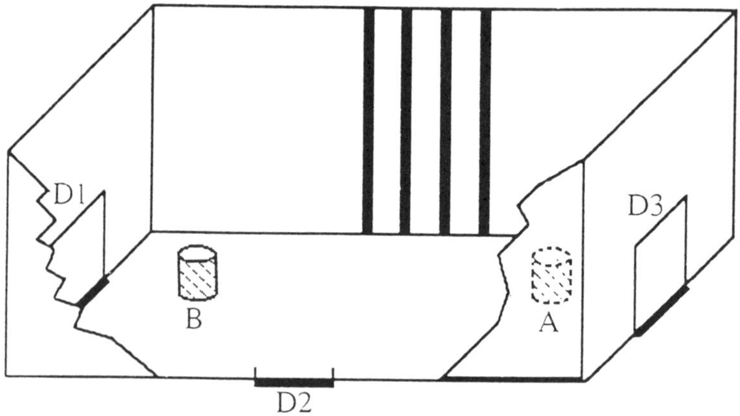

Fig. 2.4. Diagram of the apparatus with the familiar object in position B during the five trials of habituation. The "new" object identical to the familiar one is in position A in the sixth trial. (Adapted from Thinus-Blanc and Ingle, 1985.)

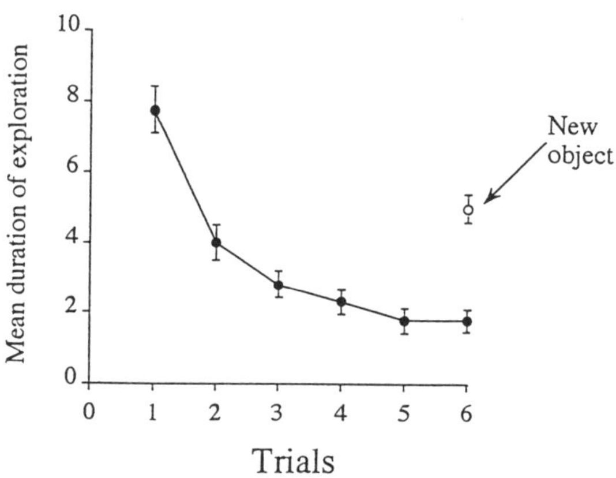

Fig. 2.5. Mean duration of exploration of the familiar object (trials 1 to 6) and of the "new" (trial 6) object. (Adapted from Thinus-Blanc and Ingle, 1985.)

heterogeneous, the change consists of uniformizing the field by removing a stimulus contained in it. Therefore, the reaction to novelty takes the form of search rather than of investigative behaviors which can be more or less focused in proximity to the locus of the missing stimulus. This procedure reduces the possibility that the animals rely on particular local views since the environment is visually very poor. For example, Corman and Shafer (1968) used a black field with a white square in the center. After a phase of familiarization, the white square was replaced by a black one. Following this stimulus change which made the field visually homogeneous, an increase of approaches to the locus of the change was recorded.

Another dishabituation test of spatial memory has been devised (Thinus-Blanc, Save, Buhot, and Poucet, 1991a; Thinus-Blanc, Save, Poucet, and Buhot, 1991b; Save, Buhot, Foreman, and Thinus-Blanc, 1992a). The general procedure was the same as in the experiments in the circular open-field discussed above but, instead of objects contained in the field, a conspecific in a small cage was located under the glass floor. This type of "stimulus" elicits strong investigative reactions which can be quantified by measuring the time spent in its proximity. After habituation had occurred, the stimulus was removed. During the test session the time spent on the locus corresponding to the previous location of the conspecific was recorded and compared to the time spent above a neutral zone arbitrarily defined by the experimenter on the video screen (see Fig. 2.6).

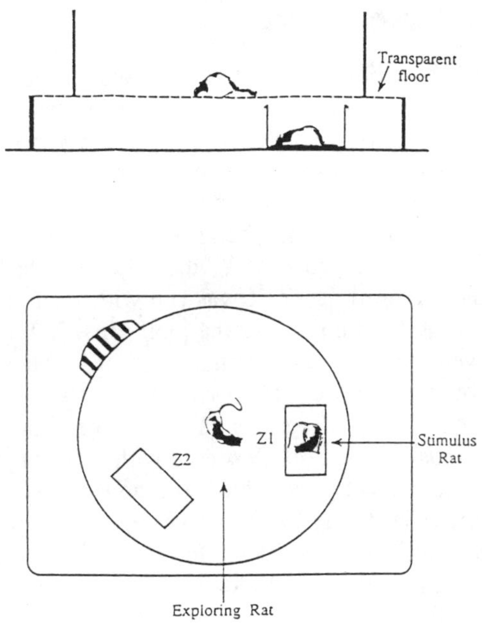

Fig. 2.6. Top: schematic representation of the apparatus with a "stimulus-rat" placed under the transparent floor; Bottom: the corresponding video image with a zone Z2, delimited on the screen, symmetrically placed to that defined by the "stimulus-rat"'s cage (Z1). During the test, the "stimulus-rat" is removed. The time spent above each of the two zones which are defined only on the video screen but not in the apparatus are recorded. (Reproduced from Thinus-Blanc et al., 1991b.)

Throughout the experiment, olfactory cues were carefully neutralized. After habituation had occurred, removing the stimulus-rat elicited an increase in the time spent above the place Z1 (where the stimulus was previously located) in comparison with the neutral zone Z2. This procedure has proven to be a useful tool for testing the effects of lesions or temporary inactivation of brain structures (see Chapters 6 and 7) although it should be noted that the reaction to a missing stimulus is not always symmetrical to orienting responses toward a newly appearing stimulus (Foreman and Thinus-Blanc, 1987).

In conclusion, these data demonstrate unambiguously that during exploratory activity spatial relationships are spontaneously encoded. Since, in most cases, the detection of the spatial change appears to be independent of the direction of approach, this means that places labelled by objects have been processed so as to allow the extraction of the invariant structure of the experimental situations.

4. Exploration and Problem Solving

Besides the dishabituation experiments, the beneficial effect of exploratory experience has been demonstrated in problem-solving tasks. An early study by Maier (1932), who was also a pioneer of cognitive psychology, is of particular interest. Rats were allowed to explore the "three-table" apparatus which consisted of three platforms and interconnecting runways (Fig. 2.7). This exploration, however, was not free but unidirectional and fragmented; one day the rats ran X-Y several times, finding food at Y; the following day, they ran Y-Z, the food being at Z, and the third day Z-X was run with food at X. During the test, the rats were shown the food on one of the platforms, X for example, and for a few seconds they were allowed to eat a small part of it. They were then released from Y. The aim of the exercise was to return directly to the table where they had just been fed by running a segment (Y-X) of the apparatus they had taken before, but in the opposite direction. A screen with a door before each table prevented the animals from seeing one table from another. Unlike a control group which had been allowed to freely explore each leg of the apparatus, rats experiencing unidirectional and fragmented exploration failed to choose the shorter path during the first test trial but ran along the longer one in the direction with which they were familiar.

In a follow-up experiment, the segments X-Z and Z-Y could be explored in both directions and in a second pre-test phase rats ran X-Y-Z for food. Only the third leg was run unidirectionally (X-Y). During the tests, rats were fed at X and placed at Y. Therefore, the optimal choice corresponded to the path Y-X which had not been explored in that direction before. Unlike the previous experiment, a large number of rats immediately chose the shorter path.

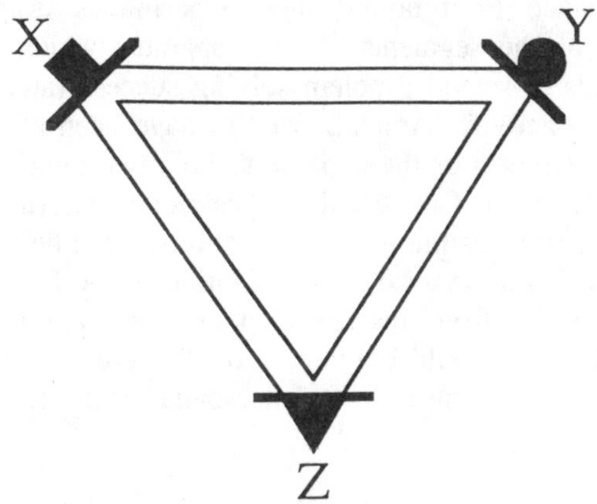

Fig. 2.7. Top view of the Maier three-table maze. A screen with a door is placed before each platform in order to prevent the rats from seeing a platform while at others.

This elegant experiment is a demonstration of the functional role of exploratory activity in the setting up of spatial relationships; it also illustrates some properties of routes and maps respectively. In the case where rats are denied bi-directional exploration, they just learn to orient according to the familiar direction, at the starting table. The structure of the maze itself determines a route once the initial choice, during the test, has been made. The behavior is dependent on the sequence in which information has been collected. In contrast, rats provided with bi-directional exploration appear to establish spatial relationships between places regardless of the temporal features of the acquisition phase. Even when bi-directional exploration did not take place on the segment of the maze used in the test, the spatial representation of the whole situation allowed the inference of the direction of the goal, i.e., discrimination of the shorter path which had hitherto not been experienced in that direction.

In a modified version of Maier's three table task with a different runway configuration (Fig. 2.8), Ellen, Soteres, and Wages (1984) showed that "piecemeal" exploration of the apparatus can lead to successful performance depending on the segments which have been separately explored. For instance, no rat that explored one table and runway per day (Fig. 2.8a) was able to solve the task whereas 60% of the rats that explored two tables and their interconnecting runways (Fig. 2.8b) did succeed. All rats that explored the entire apparatus (Fig. 2.8c) on each exploratory day were able to solve the task. It should be emphasized that the amount of exploration per se was the same for all groups. In another experiment, Ellen, Parko, Wages, Doherty, and Herrmann (1982) demonstrated that exploration of either the tables or the runways alone led to unsuccessful performance. From both of these experiments, it appears that there are the connections of the segments of the apparatus which are investigated separately which leads to spatial problem solving success rather than the mere amount of exploratory activity. Animals need to have been allowed continuous exploration of basic elements of the apparatus, i.e., two tables and the runway linking them. As a matter of fact, the three basic elements (the runways) have one part in common corresponding to their junction. It can be assumed that it is this partial overlap which allows the re-organization of the fragmented information into a whole. It is also from this central point that panoramic views involving the three possible goals could be perceived. Therefore, this central choice-point may have served as a reference for the spatial organization of the whole situation.

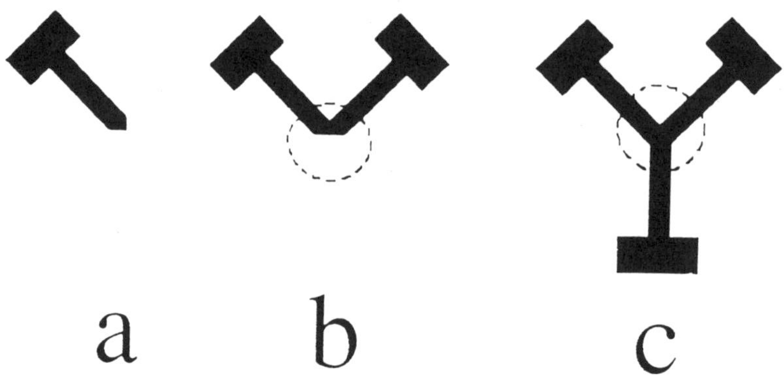

Fig. 2.8. A version of the Maier three-table maze and the various segmentations of the maze used for exploration (adapted from Ellen et al., 1984).

Indeed, results obtained by Poucet, Bolson, and Herrmann (1990) in a modified version of the maze supports this interpretation. Interconnecting systems with longer runways could be added to the initial apparatus consisting of the three tables and direct runways (Fig. 2.9). The primary aim of this experiment was to test the deficits of brain-damaged animals, though the observation of intact rats' strategies is of particular interest vis-a-vis their adaptation to the increased complexity of the situation. The groups tested in conditions a and b (Fig. 2.9) consistently chose the most direct route between tables. However, in the more complex situation c, normal rats demonstrated a strong preference for the inner route, leading to the central choice point, rather than the direct route. The main difference between conditions b (which also provided an inner runway pattern) and c is that the latter situation provided a much more complex runway pattern as compared with the condition in terms of the number of possible routes. The authors attribute the preference for the inner runways to the advantages that it provided:

"*First, it allowed delaying choice until animals were closer to either goal table than was possible by the other available routes. Second, it permitted animals to self-correct at the choice-point by gathering additional relevant information at that place. Lastly, in the configuration c, the number of alternatives at the choice-point was reduced to two routes, whereas five routes were available at the start in the same condition.*" (Poucet et al., 1990, p. 380).

When the rats became more familiar with the problem, their preferences switched from the inner route to the direct route. Together with the experiments described above concerning the effects of "piecemeal" exploratory experience, these data demonstrate that some places and their inter-relations are of particular importance with regard to spatial organization. One of these places in the Y-shape three-table apparatus appears to be the central choice-point offering views of the all three goals (including the visual background) from the same location.

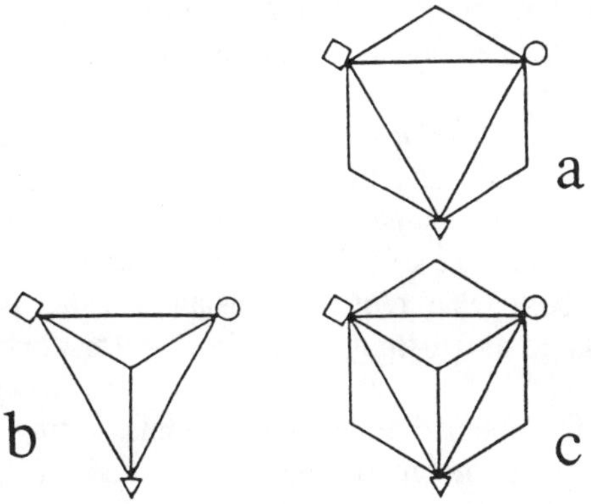

Fig. 2.9. Complex versions of the Maier three-table maze with various interconnecting runways (adapted from Poucet et al., 1990).

5. How Can Exploration Act as a Spatial "Builder"?

One of the characteristics of exploration, at least in rodents, is intense locomotor activity. In addition, it is indisputable that exploration results in some knowledge of the qualitative properties of various environmental elements; far less obvious is the fact that such behaviors are necessary in the working out of spatial relations between these elements. What is it about exploration which allows the acquisition of this specific knowledge?

Without doubt, the clearest answer to this question was given by J. J. Gibson, some decades ago. The spirit of Gibson's thoughts on this matter is summarized in the following, rather paradoxical, terms:

"*If change means to become different but not to be converted in something else, the assertion ("the more it changes, the more it is the same thing") is true and the saying emphasizes the fact that whatever is invariant is more evident with change than it would be without change. One arrangement does not become a wholly different arrangement by a displacement of viewpoint. There is no jump from one to another, only a variation of structure that serves to reveal the nonvariation of structure*" (Gibson, 1979, p. 73, edition 1986).

This assertion points to the ever changing vistas of the environment generated by all these behaviors constituting exploratory activity (locomotion, scanning movements, rearing up, and so on). For instance, the various visual images of the same arrangement of objects when perceived from different perspectives could appear as originating from various arrangements. In the simple case of four objects that form a square, i.e., a familiar arrangement, the human subject, at least, is easily able to extract the invariant square structure of the set. However, in unexpected and more unfamiliar situations, this knowledge cannot be acquired at once and displacements are required to apprehend the situation as a whole.

This dynamic feature of the process, which is the building block of spatial knowledge, provides the basic and fundamental information that very different percepts may belong to the same entity. Even when engaged in various activities, the pervasive presence of an optical flow pattern during displacements provides information about what remains constant in spite of the change. The optical flow pattern is to be distinguished from vistas opened from various perspectives.

Gibson's formulation about the spatial invariants arising from variations in the perception of the physical environment (generated by locomotor activity) finds its better instantiation through the visual modality. He coined the expression "visual kinesthesis", meaning that feedback about the movements of one's own body is not only provided by vestibular and proprioceptive information (from muscles and tendons), but perhaps more importantly, by the visual flow. It allows the easiest and most accurate distinction between perceptual changes due to one's own movement from those due to object movement. Vision ensures the stability of the external world while moving around.

It should be mentioned that Gibson provided only a description of how a stable representation is extracted under conditions of movement. Understanding the process itself poses a difficult computational problem which has not been directly addressed so far, excepted by Sharp (1991) in a model aimed at accounting for hippocampal place-cell activity (see Chapter 9). However, the spatial deficits of people who have never experienced vision (congenitally blind) or have lost sight in the early period of life, before the acquisition of autonomous mobility (early blind) are often interpreted by the authors (for example, Rieser, Guth, and Hill, 1986) along the lines of Gibson's theory. Although they do not explain in computational terms the extraction of spatial invariants, they confirm Gibson's hypothesis about the conjoint function of movement and of its result at the perceptual (visual) level.

I have described above some data (Ellen et al., 1982, 1984) showing that a global representation can be constructed only if some important junctions of the maze have been explored in continuity. Indeed, that the continuity of displacements within some regions of space is an important factor, though less crucial than visual continuity, has been demonstrated recently in an experiment by Save, Granon, Buhot, and Thinus-Blanc (1996b) using the dishabituation paradigm. Rats were allowed to explore separately and successively the two halves of a circular open field containing objects. The partition separating the two halves was either transparent or opaque allowing or preventing animals from visualizing the whole situation. In addition, locomotor access to the other half could be permitted or denied via two openings at the extremity of the partition (see Fig. 2.10). Therefore, four experimental conditions were defined: transparent and opened partition (group V/O), transparent and closed partition (group V/C), opaque and opened partition (group NV/O) and opaque and closed partition (group NV/C). Once exploration was complete, rats from these experimental groups (EXP) were exposed during the test session to the whole field without the partition. For four control groups (CNT) tested in the same situations as those presented above, the partition remained during the test session. The degree to

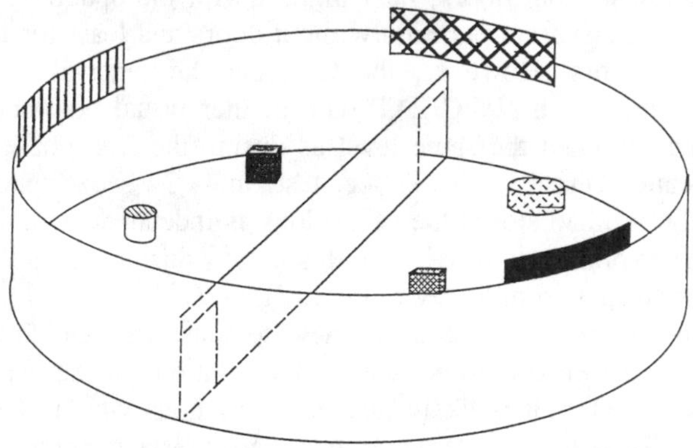

Fig. 2.10. Schema of the open-field containing four objects and an opaque or transparent partition which could be opened at each of its extremities (adapted from Save et al., 1996b).

which a unitary spatial organization had been built up was examined through the measure of reaction to novelty (contacts with the objects). As in the other dishabituation experiments, the differences between the scores recorded during the test-session and the last session before the test were computed, because it may happen that animals have different levels of overall exploratory activity but with the same variations (habituation and reaction to the change) during the course of the experiment. Whereas none of the CNT groups increased their exploratory activity, rats which had had visually discontinuous experience (i.e., with the opaque partition: NV/O/EXP and NV/C/EXP) whatever their locomotor experience (continuous or discontinuous) of the two halves of the apparatus, displayed a renewal of exploratory activity (Fig. 2.11, top). Conversely, rats having explored with the transparent partition (V/O/EXP and V/C EXP) did not react significantly to its removal (Fig. 2.11, bottom). Although a tendency was observed in the group without continuous visual and locomotor experience (NV/C/EXP) to react to the change more than the group with the discontinuous visual experience and continuous locomotor activity (NV/O/EXP), the difference did not reach statistical significance. The lack of reactivity of rats that had been provided with a continuous visual perception of the whole situation, whatever their locomotor experience, strongly suggests that spatial relationships between the four objects had already been established on the basis of visual information since removing the partition did not induce an updating of the representation already set up.

A more parsimonious hypothesis is that the more or less important re-exploration during the test corresponds to an updating of the situation, according to the importance of the change. Indeed, the exploration level was highest with the opaque partition without doors, next highest with the opaque partition with doors, next for the transparent partition without doors and least for the transparent partition with doors. However, the fact that the renewal of exploration during the test in the group NV/C/EXP with neither visual nor locomotor continuous experience reached the same level as during the first phase of exploration (first exploration phase: m = 33 sec; test: m = 32 sec) supports the idea that at least this group had stored the subregions as independent representations. Accordingly, the exploratory activity of this group during the test appears to correspond to the formation of a new representation.

However one interprets these data, they demonstrate that continuous and integrated visual experience is important for exploration. In the experiments by Ellen et al. (1982, 1984), it is likely that the continuous vision of the environment that was available to animals while they were crossing important junctions in the maze had played an important part, at least as important as continuous locomotion, in the construction of a spatial representation of the whole situation.

Indeed, the notion that perceptual changes must be matched with information about the movements which generate these changes is at the very core of the

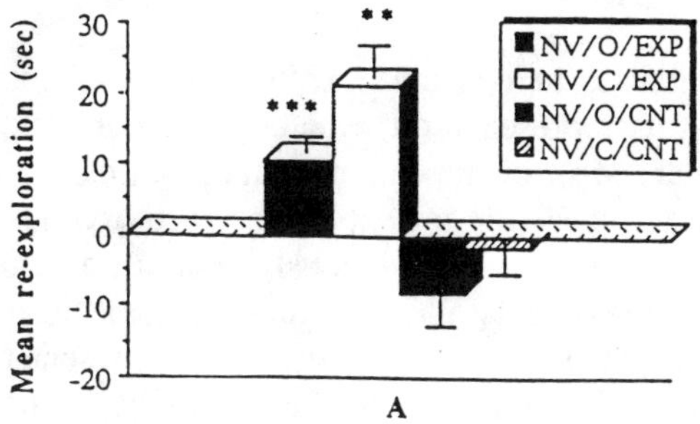

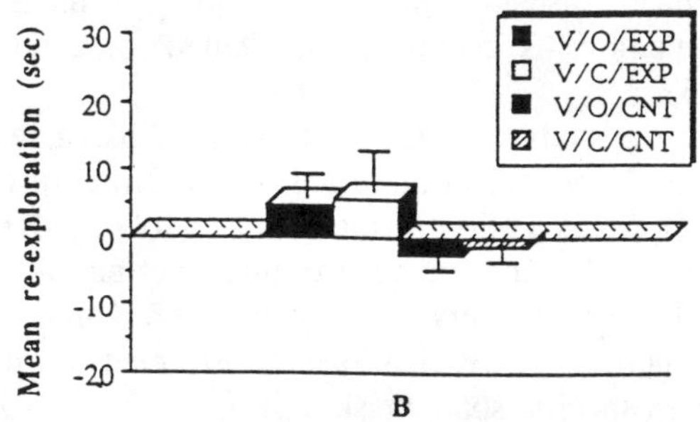

*Fig. 2.11. Mean duration of re-exploration of the objects from the last session of habituation to the test-session (removal of the barrier for the EXP groups) in the groups initially exposed to the opaque barrier (A) and to the transparent barrier (B). The stars represent the significance level (** $p<.01$; *** $p>.001$) reached by the scores of the re-exploratory activity, computed by an analysis of variance. (Adapted from Save et al., 1996b.)*

function of exploration as a "spatial builder". Exploratory activity can be considered as having both tightly inter-connected aspects: one concerning the perception of the external physical environment (perception of frontal views, for instance), the other related to self-generated internal feedbacks (movement-related information provided by muscle and tendon receptors, by the vestibular system and by the perception of the optical flow). This distinction roughly refers to what Baker (1987) has called "location-based" and "route-based" information. The latter type of information is very germane to the definition of the path-integration phenomenon which will be discussed in Chapter 5.

6. Concluding Remarks

It appears legitimate to conceive exploration as an interface between the physical world and its representation, whatever the complexity and level of abstraction of the latter. Most of the data presented in this chapter provide experimental support for a crucial role for exploratory activity. The function of exploration as a "spatial builder" is unambiguously demonstrated by a host of studies. The dishabituation experiments allow us to have some insight into the parameters which are spontaneously encoded in spatial representations. For instance, geometrical and topological relationships between objects appear to be important features for the animal (at least for hamsters). In addition, some data emphasize the predominant role of vision over other modalities, whereas others highlight the fact that not all of the parts of a given situation have equivalent status. This last aspect has important implications since it leads to speculation about a hierarchical organization of representations with some "nodes" being of particular importance. As will be discussed later, the notion of a hierarchical organization of spatial representations is a constituent element of a number of theoretical models (Chapter 5).

In conclusion, the study of exploratory activity has made a valuable contribution toward a better understanding of animal spatial representations. However, this approach (with the exception of the problem-solving experiments) concerns only the nature of the elements encoded in maps but not their functional properties. The two following chapters are aimed at examining the latter, i.e., the various means whereby representations constructed during exploration are exploited to perform specific spatial tasks.

CHAPTER 3

THE TRADITIONAL PARADIGMS

"After a brief period of relative disuse, maze testing is enjoying a strong renaissance because it is particularly adapted to addressing the more cognitive types of issues psychologists are investigating today: the structure of memory, the organization of cognitive maps, and the foraging strategies of predators searching for preys." (Olton, 1979, p. 583)

Since Tolman cast the cognitive map hypothesis, many studies have been conducted, especially since the nineteen-seventies, in order to provide it with experimental support. As in Tolman's experiments, most of the situations aimed at investigating the nature of spatial representations in rodents are multi-trial learning tasks requiring the subjects to locate and reach a generally invisible goal in a maze or an arena. Indeed, the animals must be motivated enough to complete the task. All the paradigms involve, therefore, a reinforcing event encouraging animals to implement a proficient strategy for reaching the goal either because they will find a reward there, or, alternatively, a safe place to escape from an uncomfortable medium (such as from cold water in the Morris swimming pool). Thus, these situations differ markedly from those described in the preceding chapter, which generally take place within a relatively short period of time, rely on spontaneous behavior, and (with the exception of the problem-solving tests) do not involve any explicit reinforcement. The fact that animals are repeatedly exposed to the same situation also represents a consistent difference with the formerly described paradigms. Animals subjected to a training experience may easily identify, trial after trial, the spatial relationships existing among some locations – the starting point, the choice points or the goal – which provide different views of the experimental layout, and have a particular status in the picking up of information.

It is therefore apparent that analyzing the properties of spatial representations in learning situations implies that the observed behavioral strategies will always depend on a balance between the animals' cognitive abilities and the task requirements.

The term "strategy" has a wide range of meanings, depending on the context in which it is used. In most of the studies that are discussed in the present chapter, the authors seldom provide their own definition of "strategy". The term, however, can be simply taken as referring to a set of functional rules which are implemented by the animal at the various phases of information processing. This set of rules is supposed to allow the animal to reach an acceptable performance level without excessive cognitive effort. Consequently, this notion refers to some kind of compromise, i.e., to an optimal ratio between cognitive load and efficiency. In this respect, the definition of "strategies" that I propose here is very close to the main principle underlying foraging theory (i.e., maximized energy gain and minimized expenditure of energy and time; see Krebs and McCleery, 1984; Stephens and Krebs, 1986) which is discussed in the following section.

The strategies that animals adopt, and which vary between individuals, are more likely to be observable when they are performing tasks with a high degree of difficulty, or when it is necessary to compensate for deficits, than when the problem is easy to solve. In the latter case, a minimal cognitive effort permits the subject to obtain the maximal reward, so that the notion of "compromise" or optimal ratio between cognitive load and efficiency becomes irrelevant.

Most of the studies presented in the present chapter were aimed at demonstrating the existence and properties of spatial representations or cognitive maps. However, according to the situation, the use of maps, reflecting high-level cognitive processing, may not be the most economical strategy to implement. For instance, "guidances" and "orientations-directions" proposed by O'Keefe and Nadel (1978; Chapter 1) provide examples of fast and simple means of navigating toward a goal. Other types of strategy have been put forward to account for the accurate localization of an inconspicuous goal.

One of them is the "Snapshot Memory" hypothesis proposed by Cartwright and Collett (1982, 1983). Although emanating from studies in bees, this type of orienting mechanism is sometimes applied to rodents. The hypothesis states that the localization of an inconspicuous goal relies on a process of comparison between a stored image or snapshot of the landmark(s) as seen from the goal and the current image of the environment. Attaining the goal would rely on obtaining the best match between the stored and current retinal images.

Another possibility, this time based on studies in gerbils, has been proposed by Collett and his collaborators to account for the localization of a hidden goal. It concerns the "Vector Sum" model (Collett, Cartwright, and Smith, 1986). On the basis of experimental data, the authors hypothesize the existence of two vectors, which seem necessary for the animal to localize a desired place before planning a trajectory: a "seen" vector corresponding to the animal's position with respect to the landmark and a "stored" vector which gives the position of the goal with respect to that landmark. The animal can thus compute a direct trajectory to the goal by calculating the difference between the two (see Fig. 5.1) even from a new direction of approach (the Vector Sum hypothesis is discussed at length in Chapter 5).

In this chapter, the experimental paradigms which have been the most frequently used to study animal spatial cognition, and the main issues which prompted their creation, are reviewed. Besides the two popular situations, the Olton radial maze and the Morris water maze, others, such as the cross-maze, have also contributed to progress in this domain. Another aim of this chapter is to critically examine the contribution of these studies keeping in mind that means of orientation other than maps can account for spatial learning performance. In this regard, three questions will be considered: 1) Do the experimental data demonstrate the existence of cognitive maps as defined by Tolman and by O'Keefe and Nadel? 2) Can more parsimonious explanations, such as those evoked above, account for these data? and 3) Are new properties of spatial representations emerging from the more recent paradigms?

1. Foraging Behaviors

The "multiple solution problems", including the Olton radial maze, which are presented in the following sections represent to some extent the experimental counterpart of situations frequently encountered in natural conditions. In this respect, a preliminary short survey of ethological studies in this domain is appropriate since it allows us to underline the interest and power of the multiple solution problems. These are not experimental situations unrelated to natural behaviors, but they do tally with a spontaneous tendency which is observed in many species. This tendency consists of not making immediate revisits to places where food has just been obtained. Such "win-shift" behaviors are usually contrasted with "win-stay" behaviors (consisting of returning to the same food source). Obviously, these habits greatly depend on the distribution of food resources in natural conditions, i.e., whether small amounts of food are scattered over large areas of space favoring win-shift – or there is a single limitless source of nutriment favoring win-stay – (see also Haig, Rawlins, Olton, Mead, and Taylor, 1983).

Several species hoard small amounts of food in a scattered distribution within their home range or territory and are able to recover their caches. The whole process (hoarding and foraging) can take place within one or more days. The animals generally store food in a season when it is available and recover it during the winter when it is in short supply (see Gallistel, 1990, for a review). This particular spatial ability appears to be shared by many species (Olton, 1979) even if different strategies are used (Shettleworth and Krebs, 1986). Foraging behavior has been the subject of several theoretical considerations. One statement of the foraging theory is that animals have been genetically programmed to make decisions that will tend to maximize energy gain and minimize expenditure of energy and time (Krebs and McCleery, 1984; Stephens and Krebs, 1986).

Studies of spontaneous behavior in natural or semi-natural conditions have revealed some important features of this remarkable example of animal spatial memory. From observations in the wild and laboratory experiments, Sherry (1987) has obtained evidence for some of the underlying mechanisms in this process. For instance, Cowie, Krebs, and Sherry (1981) have demonstrated that the recovery of stored food does not occur by chance. Birds (marsh tits) had access to a feeder containing radioactively labelled sunflower seeds. After the bird had taken a limited number of them, the experimenter located stored seeds by means of a scintillation counter. Then, two control seeds were placed near the radioactive seed. After several hours, the sites were inspected. It appeared that most of the seeds stored by the birds had been removed whereas many of those hidden by the experimenter were still in place. This experimental procedure allows us to rule out the possibility that the stored seeds had been pilfered by other animals and that the caches had been re-encountered by chance by the food-storer.

Other experiments conducted in large indoor aviaries with marsh tits and black-capped chickadees have unequivocally established that visual distal cues are used to locate the seeds and that the route used during recovery is different from that used while establishing caches (Sherry, Krebs, and Cowie, 1981; Sherry, 1982, 1984). In addition, if they are given two qualities of seeds, one of them being preferred, the birds visit only the caches containing the preferred category of seed. As a matter of fact, this distinction represents a type of chunking process similar to that which is referred to in Section 4.

Although they do not explicitly refer to foraging, the behaviors observed by Menzel (1978) in young chimpanzees represent another related example of the use of spatial memory in retracing different food caches. Pieces of food were hidden by the experimenter at 18 different points in an open outdoor field. The experimenter carried the animal around the field so that it could observe where the food was located. Then the animal was let free to collect the hidden pieces of food. The chimps recovered about 12 of the 18 pieces of food by the end of the trial. Interestingly, they collected food regardless of the order in which it had been hidden by the experimenter but used a least-effort principle, i.e., they collected all of the food in one area before moving to another area. Similar data have been reported by Vauclair (1990) in baboons.

These data, including those where the animal itself has hidden the food, share common processes. In every cases, short term memory is necessary to update the situation as a function of the subject's action so as not to avoid already emptied caches to be revisited, while long term memory is required for determining the cache locations.

It should be emphasized that birds' behavior represents an unusual case by comparison with terrestrial animals since some bird species are able to retrieve food hidden in as many as several hundred places. This extraordinary capability may be related to the fact that birds are, ipso facto, provided with survey views of the environment; they can match frontal views when, for instance, they skim along the ground. In contrast, terrestrial animals are not provided with any survey view of their environment. That may be one of the reasons why the spatial memory capacity of rodents, felidae or primates appears to be less efficient than that of birds.

2. "Multiple Solution" Problems: the Precursors of the Radial Arm Maze

The distinctive feature of "multiple solution problems" or "multiple choice situations" that differentiate them from other types of maze (cross maze, water maze, etc.) is that successful performance is not associated with the adoption of a single trajectory leading to a unique goal. Usually, multiple choice situations include several baited places that the animal is free to visit in any order until all the food has been collected. In this respect, these problems can be considered as experimental equivalents of foraging in natural situations. The large amount of freedom that is afforded the animal in terms of behavior may enable the identification of some principles of its behavioral organization and the implementation of strategies, via the analysis of patterns of visits.

The first multiple solution problem to be used was the "progressive elimination of food" or simply "elimination" problem. It was first developed by Crannell (1942). The animal was placed in a maze consisting of at least three alleys diverging from a common starting platform. A piece of food was placed at the end of each alley, and the subject had to learn to find this reward without returning to a previously visited (and hence empty) alley. Several animal species are able to solve this problem: rats (Lachman and Brown, 1957; Lachman, 1965, 1966, 1969a and b), golden hamsters (Buhot and Teule, 1971; Teule, Buhot, and Durup, 1972; Ammassari-Teule and Durup, 1982), cats (Poucet, Buhot-Averseng, and Thinus-Blanc, 1983a) and dogs (Fabrigoule, 1974). For the latter two species, the experiment took place in an open area (outdoors, or in a large room) with one starting place and with feeding bowls instead of maze alleys (Fig. 3.1). The number of trials required to reach criterion (a trial corresponding to the series of visits required to successfully collect all the food) varied depending on both species and experimental conditions. Despite this variability, a common feature which was found to emerge was that animals generally showed, during trials in which no errors were committed, a tendency to make successive choices to alleys that were maximally divergent. The degree of divergence was measured by the sum of the spatial intervals between successively chosen bowls during a single session (Table 1). Data showing the preference for high divergence patterns of choice in cats (Poucet et al., 1983a) are presented as an example in Fig. 3.2.

Table 1.
Example of a pattern corresponding to the highest degree of divergence (7) for a situation with four goals. The pattern of choices here is "2-4-1-3". (Adapted from Buhot and Poucet, 1987.)

Sequence of visits	Goal numbers 1	2*	3	4	Number of divergence units
1st to 2nd		|———	———	———↑	+ 2
2nd to 3rd	↑———	———	———	———|	+ 3
3rd to 4th	|———	———	———↑		+ 2
					S = 7

* = first chosen.

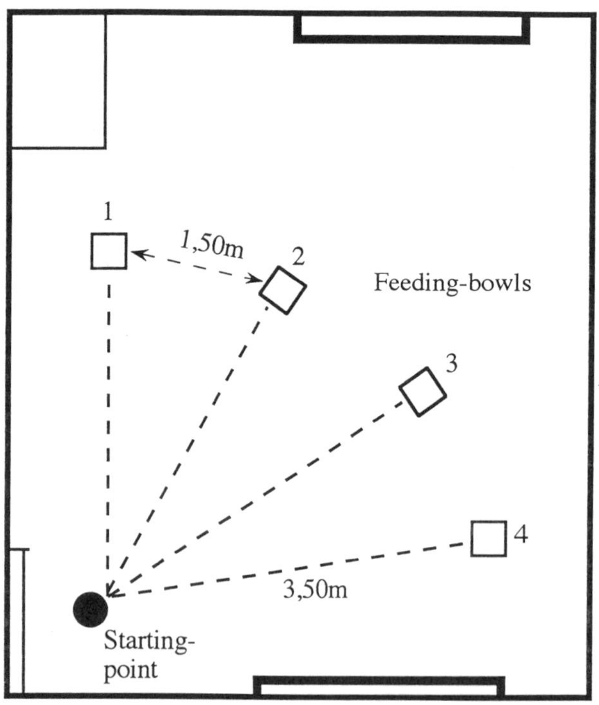

Fig. 3.1. Layout of the experimental room with the location of the four feeding bowls (represented by squares) used in the experiment with cats. (Adapted from Poucet et al, 1983a.)

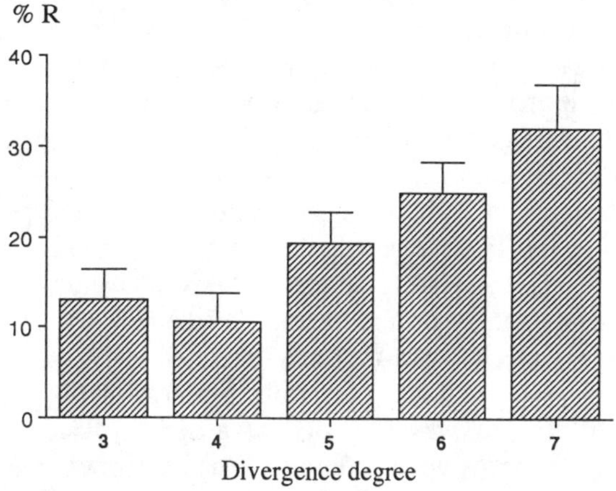

Fig. 3.2. Mean distribution of the elimination patterns for trials without errors, for each divergence degree category (adapted from Poucet et al., 1983a).

This phenomenon is closely related to the tendency to alternate (spontaneous alternation) that had already been observed in two-path mazes (see Chapter 2) in the absence of reinforcement. The latter data were interpreted in terms of "stimulus satiation" (Glanzer, 1953), or curiosity and the search for novel stimuli (Dember, 1956). In mazes involving more paths, it is apparent that searching for food requires a more consistent processing of spatial information to be performed. Lachman and Brown (1957) tested rats in a three- or a four-arm radial maze and observed that, in the last training trials, the animals tended to run the most divergent paths between choosing the central ones. Buhot and Teule (1971) observed a similar behavior in hamsters and concluded that this "maximum divergence principle" might be a strategy for making two correct runs immediately – and thus obtaining food twice consecutively – since the most divergent paths correspond to the best differentiated spatial locations. This type of strategy is not the simplest since it relies on the storage of discrete events and cannot be easily applied when the number of arms is increased. Conversely, the simplest win-shift strategy, which rapidly produces a high level of performance, consists of running to adjacent arms. It is worth noting that implementing this simple algorithm does not require any spatial representation of the maze since on each choice, animals have just to memorize the direction (clockwise or anticlockwise) of the previous visit.

Successively, Buhot and Teule (1971) and Ammassari-Teule and Durup (1982) attempted to analyze more accurately the influence of the site of reward on performance. They used a modified version of the original situation, i.e., a spindle-shape maze with three arms diverging from a common platform which then converged to a common goal box (Fig. 3.3). As before, the task for the hamster was to run each path once only and collect all three food pellets. The particular shape of the maze, however, allowed the experimenters to place the food cups at different locations: 1) at the halfway point of each arm, i.e., in three well-differentiated locations, 2) at the end of each arm, i.e., in three close but still distinct locations, or 3) in a single location, i.e., at the point of convergence of the arms. In the latter case, there was a single food cup that was replenished following each first visit to an arm. The results showed that hamsters' performance depended strongly on the location of the reward. Performance was better (i.e., the number of choices of already-travelled paths was reduced) when the reward locations were separated (situations a and b in Fig. 3.3) than when the same amount of reward was provided at a single place. Moreover, the sequences of visits to the arms were different. When the reward was provided at a single place, hamsters implemented the simple lateral scanning strategy described above, i.e., they visited the arms systematically from left to right, or vice-versa. In the other situations, no stereotyped patterns of visits were observed and the order of choices conformed to the "maximum divergence" principle.

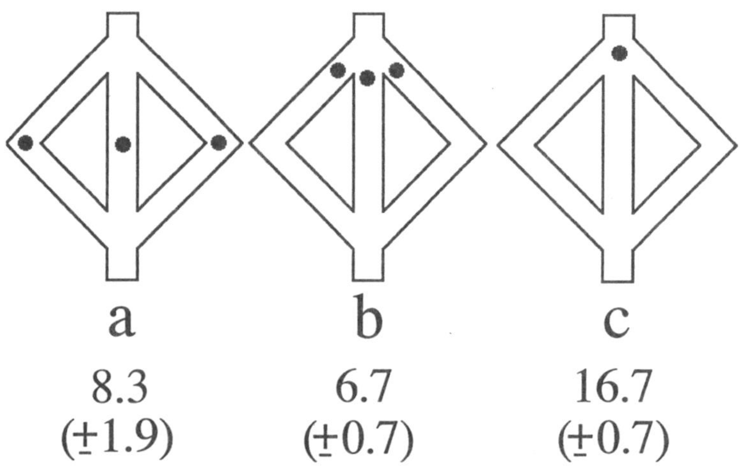

Fig. 3.3. Schemas of the spindle apparatus with the food placed at different locations and performance levels (adapted from Buhot and Teule, 1971).

In another series of experiments based on the same principle, Buhot and Poucet (1987) demonstrated the strong influence of maze structure on the patterns of visits made by hamsters. When the structure was complex (with ramifications), or when the goals were close to each other, stereotyped and simple strategies were implemented to obtain the reward. Conversely, highly divergent choice patterns were found when the structure of the maze was simple and the goals were far apart from one another.

These experiments stress an important distinction, already discussed in Chapter 1, between two main constituent elements of the orientation process: localization of the goals and choice of the path leading to them. The fact that the goals may or may not be uniquely spatially differentiated by a specific pathway leading to each of them appears to be an important feature of the situation for spatial encoding, at least as important as the structure of the maze.

3. Olton's Radial Arm Maze

In this popular maze, several paths (usually eight, but sometimes as many as seventeen) radiate from a central platform. A small amount of food is hidden at the distal end of each of the arms, in such a way that it cannot be seen by the subjects from the central platform. On each trial, the animal is placed in the apparatus on the central platform and left free to visit the arms and eat the food. Learning criterion is reached when the rat displays an optimal pattern of food collecting, with a minimal number of visits, i.e., visiting all arms without entering previously visited (empty) arms.

This test is very germane to the "progressive elimination of food" problem discussed above since in both situations the task consists of collecting food in several places. Differences lie in the fact that animals are not withdrawn from the radial maze after each choice, that the structure of the maze is circular and the maze itself provides no information about orientation. In addition, there are often more items to be processed than in the other mazes.

This spatial problem and several of its variants have been successfully used not only in rats (Diez-Chamizo, Sterio, and Mackintosh, 1985; Poucet, Lucchessi, and Thinus-Blanc, 1991b; Suzuki, Augerinos, and Black, 1980, for instance) but also in hamsters (Jones, McGhee, and Wilkie, 1990), mice (Ammassari-Teule and Caprioli, 1985; Pico and Davis, 1984; Rossi-Arnaud and Ammassari-Teule, 1994; Rossi-Arnaud, Fagioli, and Ammassari-Teule, 1991), tits (Hilton and Krebs, 1991), pigeons (Honig, 1978), doves (Wilkie, Spetch, and Chew, 1981), fish (Roitblat, Tham, and Golub, 1982), dogs (Fabrigoule, 1987), snakes (Kunka, Bernstein, and Kubie, 1979), and children (Aadland, Beatty, and Maki, 1985; Foreman, Arber, and Savage, 1984).

Using the original version of the maze with eight baited arms, Olton and collaborators (e.g., Olton, 1978; Olton, Collison, and Werz, 1977; Olton and Samuelson, 1976) reported that rats do not use stereotyped patterns of visits over trials, nor intra-maze cues such as olfactory scents, nor algorithmic strategies such as always entering the alley adjacent to the one just exited[a] and proceeding around the maze in a clockwise or counterclockwise direction (Olton and Collison, 1979). However, motor strategies have been found to interfere with more "cognitive" behaviors in rats (Foreman, 1985). That may explain why a factor which affects performance accuracy and the pattern of visits is maze regularity, i.e., the fact that the arms are regularly spaced. Grobéty and Schenk (1992) trained rats in an irregular radial maze in which the angle between adjacent arms varied between 10° and 90°. In that case, the most frequent pattern of visits chained the five arms that divided space most regularly and rats neglected the remaining three arms. As a consequence, performance was less accurate than in a regular maze in the same environment. The authors concluded that "... the representation used for patrolling was, at least partially, determined by the maze's geometrical properties" (Schenk, Grobéty, Lavenex, and Lipp, 1995, p. 291).

These data stand in contrast to those obtained by Buresova (1980) who used a non-radial multiple goal maze. She found that the acquisition rate was almost the same as in the radial maze. The difference between these results and those by Grobéty and Schenk may be due to the fact that, in Buresova's experiments, tubular segments equipped with one-way entrance and exit doors were scattered in an open-field. Thus, rats had to visit all elements of this "random maze" in order to collect a piece of food from each segment without having to return to a central zone.

The shape of the maze arms was also found to play a role in choice accuracy. Usually, the arms are straight and the initial direction to be taken from the central platform is confounded with the direction of the goal. Using a modified version of the radial maze with angled arms which required a right turn at the end of each, Poucet et al. (1991b) have differentiated the location of the goals from the initial direction to be taken. They have shown that rats make complementary use of both types of information. However, during transfer tests between angled and straight arm conditions, the increase in errors was found to be much more pronounced when the initial location of the goal was modified than when rats had to use new travel directions. This result strengthens the idea that the localization of places (the goals) predominates over the routes leading to them.

Other spatial features of the maze influence the patrolling sequence. For instance, a high frequency of adjacent transitions is observed when the central

---

[a] This finding cannot be generalized to all rodent species. For instance, Ammassari-Teule, Hoffmann, and Rossi-Arnaud (1992) have shown that, unlike rats, mice increase their choice of adjacent arms with training in the 8-arm maze.

choice area is large (Yoerg and Kamil, 1982; see Schenk, Contant, and Grobéty, 1990, and Foreman, 1996, for further discussions).

Finally, factors unrelated to the spatial features of the test situation can also modify the patterns of visits. For instance high motivation (strict diet or water-deprivation) increases the probability of making adjacent-arm visits (Dale and Roberts, 1986; Innis and Macgillivray, 1987).

4. Memory Processes and Use of Landmarks in the Radial Arm Maze

Several studies have revealed changes in the mechanisms controlling choices over the course of each trial (Brown and Cook, 1986; Brown, Wheeler, and Riley, 1989). The sequence of choice in the radial maze is a dynamic process based on two types of memory (Cook, Brown, and Riley, 1985): retrospective memory (i.e., memory for previously chosen arms) and prospective memory (i.e., memory for anticipated choices). A discussion of this issue is beyond the scope of this book (see Foreman, 1996; Roberts, 1984, for reviews). In contrast, another distinction among memory processes is related to the nature of information used while navigating in the radial maze. It concerns Working Memory (WM) versus Reference Memory (RM). According to Olton and to many other researchers who have used this apparatus, short term WM allows the subject to update the situation as it is modified by its own behavior (the above mentioned retrospective and prospective memory belong to this first type). A high performance level implies that at any one time animals remember which arms have already been visited and/or which arms are still baited. Two prerequisites are necessary for this accurate distinction to occur: 1) animals have to discover and implement a general rule (not to enter an already visited arm) and 2) the position of each arm must be identified with regard to the environmental layout since the arms cannot be differentiated on the basis of their physical properties. Storage of these features of the task, which do not vary across trials, involves long term RM.

A useful variant of the standard radial maze test procedure allows the simultaneous evaluation of WM and RM. This consists of baiting only half of the maze arms (Fig. 3.4). Then, two distinct subsets of maze arms (a baited and an unbaited subset) are created. It is therefore possible to dissociate WM errors (revisiting a previously visited baited arm) from RM errors (entering an arm within the unbaited subset).

This distinction, however, is not as clear as it appears at first glance. For instance, unbaited arms can be entered several times within the same trial. How can such errors be labelled? Do they result from the failure of one or both types of memory? More generally, it is worth mentioning that the term "Working

Memory" launched by Baddeley (1986) may be misleading when applied to animals since it refers to human memory processes tightly related to verbalization and semantics; moreover, it is never regarded as being in "opposition" to RM processes (Baddeley, 1993). The distinction between short-term and long-term memory does not involve such ambiguities and is sufficiently functional for the purpose of discussing animals' behavior in the radial arm maze.

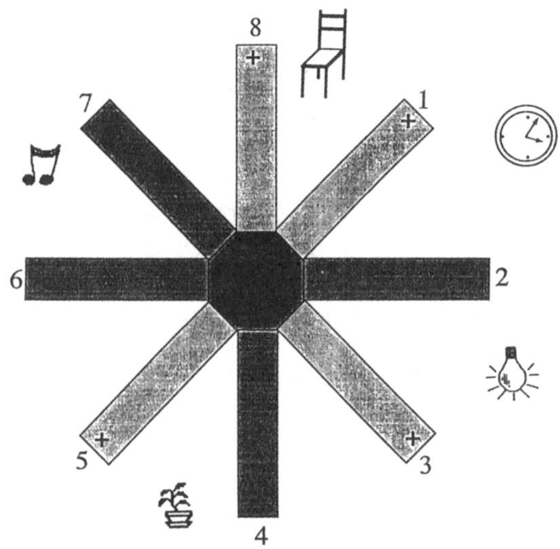

Fig. 3.4. The radial arm maze with a variety of environmental cues. Half of the arms are baited (+). The two sets of baited and unbaited arms constitute two imbricated subspaces.

The problem raised by the distinction between WM and RM is closely related to the question concerning the environmental information which is used to solve the task. Olton's initial viewpoint (1978) was that one landmark would be associated with each arm and a list of non-spatial individual items would be stored in WM. Even if he admitted that the rats were able to establish relationships between extra-apparatus landmarks, Olton argued that: "... they don't have to do so because the task can be successfully performed by treating each arm as an isolated stimulus, completely independent of each of the others" (Olton, 1978, p. 356).

Suzuki et al. (1980) tested Olton's hypothesis by placing a different conspicuous visual landmark at the end of each of the eight arms of a radial maze. After the rats had visited three of the baited arms, the relative positions of the landmarks were changed. Therefore, if the rats processed the various arms independently, one from another, they should have continued to visit the remaining baited arms according to the associated landmarks. Conversely, if they relied on the relationships between landmarks, their performance would have been disrupted. Indeed, the observed behavior confirmed the second alternative. This modification greatly affected performance, unlike a rotation of the landmarks which left their relative positions unchanged. The authors concluded that rats rely on a whole landmark configuration to differentiate the empty arms from the baited ones, as the visits progress.

A study by Mazmanian and Roberts (1983) brings additional arguments to support Suzuki et al.'s conclusions. It is based on the hypothesis that if only a list of cues (one associated with each arm) were required for efficient behavior, then preventing rats from viewing the entire environmental layout should not affect their performance. In a four-arm radial maze, rats were first placed at the ends of two of the arms and allowed to view different amounts of the surrounding space, and for different lengths of time (Fig. 3.5). They were then released from the central platform. A correct response consisted of visiting the arms on which they had not been placed during the previous phase. Fig. 3.6 shows that the accuracy of choices increased steadily as degree of spatial view widened from tunnel to 180°, to 360°. In addition, 20-sec exposure to the environment was found to lead to better performance than a 2-sec exposure for all three viewing conditions.

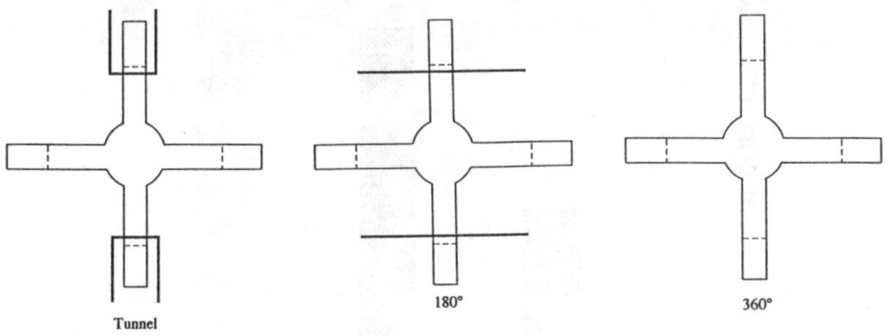

Fig. 3.5. Diagrams of the three viewing conditions used on the four-arm radial maze (reproduced from Mazmanian and Roberts, 1983).

In spite of these weighty arguments against Olton's "list" hypothesis, Brown and Cook (1986) have analyzed rats' behavior in the central arena of a 12-arm radial maze and interpreted it along the lines of Olton's hypothesis. They observed that, frequently, rats actively examined the entrance to the arms even if they had already been visited. This behavior was taken by the authors to reflect the need to be exposed to the visual stimuli associated with each arm in order to discriminate its status (previously visited, or unvisited). Brown (1992) made a more systematic study of these visual orientation movements toward maze arms, which he called microchoices (Brown et al., 1989). When micro-choices were followed by the arm being visited, they were called macrochoices. Brown assumed that microchoices constituted go-no-go decisions regarding each arm. The analysis revealed that many microchoices were not followed by macro-choices, i.e., that rats visually oriented toward many arms that they did not visit. In addition, the number of microchoices preceding each correct macrochoice increased over the course of the choice sequence. Brown interpreted these data as supporting Olton's "list" theory which states that arms are investigated one at a time and discriminated as previously visited or unvisited on the basis of information about individual arms that is stored in working memory. This theory does predict the observed increase in microchoices over trials, since there is a decrease in the density of baited arms as the choice progresses, and consequently an increasing number of arms need to be investigated before a baited arm is located.

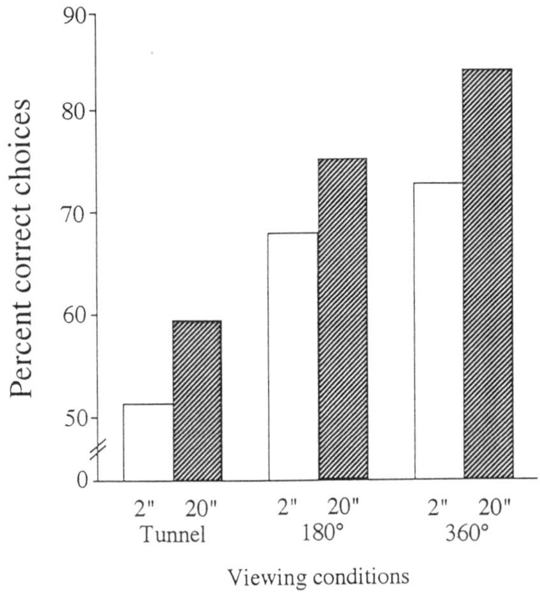

Fig. 3.6. Percentage of correct choices made under each viewing conditions, for 2- and 20-sec viewing times (adapted from Mazmanian and Roberts, 1983).

To strengthen his conclusion, Brown (1992) made a Monte-Carlo simulation of radial arm maze performance based on the assumption of the list view (that arms of the maze are individually discriminated as previously visited or unvisited). The model performed, at worst, only slightly less well than the rats in the experiments.

Brown's demonstration and model lead to opposite conclusions from those drawn by Suzuki et al. (1980) and Mazmanian and Roberts (1983). All of Brown's conclusions, however, rely on the assumption that visual orientation movements and hesitations reflect the individual processing of each arm. This does not mean that environmental cues which allow the animal to identify the status (previously visited, or unvisited) of the arms are also processed separately.

Incidentally, it is interesting to note that Tolman (1948) had interpreted such hesitation Behaviors (what he termed "Vicarious Trial and Error"), displayed by rats when they have to make a choice, as proof that information is actively selected and used to construct cognitive maps (see Chapter 1)! However, the context was so different during the Hull-Tolman controversy that the argument does not carry much weight in the present discussion. More importantly, as will be discussed later on in this book (Chapter 6), the fact that hippocampal lesions which selectively impair spatial learning have deleterious effects on radial maze performance tips the scales toward an interpretation in terms of spatial representations rather than the storage of a list of non spatial items.

All the data mentioned above demonstrate that provided neither stereotypies nor non-spatial algorithms are observed, collecting food in the radial arm maze without revisiting an empty arm is a genuinely spatial task. It requires the use of widely distributed extra-apparatus landmark configurations. This means that the status of the goals (previously visited, or unvisited) must be defined according to their location in the environment and by reference to landmark configurations. That can be done by deducing the status of the arm from the landmark configuration (along the Vector Sum model), i.e., from visual scenes perceived from the central platform. This possibility is all the more likely in that there is only one choice-place. Even if wide landmark configurations are used, the data obtained in the radial maze do not demonstrate the existence of survey maps as has often been concluded. For instance, in Suzuki et al.'s and Mazmanian and Roberts' experiments, rats did not use one particular cue associated with each of the goals but they may have relied on a wide visual scene containing some of the cues, some perhaps perceived from adjacent arms. Therefore, the modification of the visual scene associated with each of the goals, resulting from changes in the landmark configuration could account for the increased errors in Suzuki et al.'s experiment.

5. Chunking Experiments

The version of the radial maze task that involves selectively baiting half of the arms has been described in the preceding section. In that case, the whole situation is partitioned into two related subspaces (two distinct sets of baited, and unbaited arms) which can be discriminated only on the basis of whether or not they have yielded reinforcement. Therefore, reaching a learning criterion means that the animal has been able to learn and use this distinction in order to avoid the empty arms. In other words, it is capable of displaying an adaptive behavior based on classification. This represents a simple example of the so-called "chunking" process. Other examples of adaptive behavior in complex spatial tasks are provided in this section, which demonstrate the capabilities of spatial classification and serial order in animals.

Using a 12-arm radial maze, Dallal and Meck (1990) have shown that rats organize the collection of different food types in a hierarchical manner. Three types of food were each located on four arms of the maze. Provided that this baiting arrangement remained stable throughout the trials, rats subjected to this training displayed increased accuracy in choice performance and hierarchically ordered patterns of choice selection, by comparison with rats trained with the 12 arms all baited with the same type of food. In addition, if after reaching the learning criterion the experiment was transferred to a different environment, the animals trained with the three types of food needed fewer visits to collect the food than subjects in other groups. This result demonstrates that the classification that was made initially facilitated learning in a new environment. Moreover, the authors suggest that chunking may have involved the use of geometric relationships among goals baited with different food types, without reference to other extra-maze cues.

In a more analytical experiment, Ilersich, Mazmanian, and Roberts (1988) examined the performance of rats in a maze with four arms containing four "feeding stations" with different amounts of food. One group was tested with open feeders whereas for the other group, there were covers on the feeders requiring more time to gain access to the food (Fig. 3.7). In the latter case, the rats often visited the feeders containing the larger amounts of food first whereas with open feeders, they visited each of them in the order they were encountered while running from the center of the maze to the end of the arms. In addition, when tested with covered and open feeders on alternate sessions, the rats were able to switch readily between these two strategies. Thus, a chunking process may have occurred when the feeders were covered. In that case, chunking should consist of the division of the experimental layout into four subspaces, each made up of the set of feeding stations containing the same amount of food.

Chunking strategies can also be induced by using particular maze configurations. Even if visiting all the food sources in one cluster before visiting those in other clusters most likely improves rats' performance, some long-lasting neglects can also be observed in mazes with a particular structure, for example when the maze is composed of 6 radial and 2 parallel arms (Schenk et al. 1990).

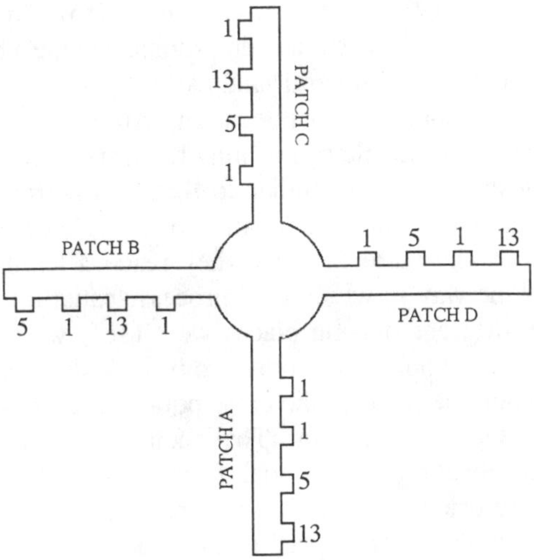

Fig. 3.7. Top view of the apparatus with feeding stations containing various amounts of food indicated by the numbers (adapted from Ilersich et al., 1988).

Such data support the existence of chunking capabilities in rats, permitting them to organize information in a hierarchical manner so as to increase the capacity of memory. In the above situations chunking could be described more concretely by the constitution and separate processing of several hierarchically related subspaces. Each subspace would be composed of items corresponding to one type or one amount of food, or to a cluster. These subspaces would be easier to process in short-term memory because of the small number of items contained in each of them. They would also constitute simpler entities (since they have fewer elements) for the purpose of storage in long-term memory.

6. The Cross Maze: a Test for Place Hypothesis?

The cross-maze consists of four arms with no distinctive intra-apparatus cues. The end of one of these arms is the goal (the reward is hidden) and the other three serve as starting points. Since the goal arm is identical to the others, the animal which is released in a pseudo-random order from any of the three remaining arms has to deduce the goal location from the available extra-apparatus cues.

This experimental design has been used in hamsters (Poucet and Scotto, 1987), rats (O'Keefe and Conway, 1980; O'Keefe and Speakman, 1987; Soffié, Bronchart, and Lebailly, 1986) and cats (Poucet, 1985). All of these experiments were conducted in cue-controlled environments: the apparatus was surrounded by curtains and one or several landmark(s) placed within the curtained area had to be used to compute goal location. All the species tested in this apparatus were found to be capable of reaching the learning criterion.

Thus, the conclusion is rather similar to that drawn from data obtained in the radial maze: animals can infer the location of a hidden goal by using visual landmark(s). However, the mere fact that they reach a learning criterion does not allow us to infer the way in which they process spatial information. In addition, although three different starting places were used, we cannot assume that animals computed correct choice directions from these different places. A more economical means would be to go to the cross point and to make the choice from this place whatever the starting arm. The choice could be made either by actually inferring the location of the goal from that of the cue (Vector Sum hypothesis), or by a re-orientation (Taxon strategy; see Chapter 1) consisting of making rotations within egocentric space so as to orient the body axis in a specific direction relative to one landmark (for instance, in anthropomorphic terms: "I should go straight ahead once the cue is behind me on the left").

One way to get some insight into this issue is first to determine at what moment in the trial the relevant information is picked up and/or the choice of the path leading to the goal is made. Delayed Response (DR) tests are adequate for this purpose: the subject is shown the situation from a starting point but is allowed to choose which route to take to reach the goal after a delay. In the meantime, the visual information relevant to locate the goal has been removed. Either the animal makes its decision from the starting places, in which case it should succeed in DR tests, or it chooses at the central cross point in which case it should fail, at least during the first DR tests.

With the aim of testing the effects of hippocampal lesions in rats, O'Keefe and Conway (1980) conducted an experiment using a cross-maze, which is also of interest at the psychological level. In their study, rats were trained to a learning criterion and then subjected to DR tests. One group of subjects performed in a situation with cues distributed all around the cue-controlled environment (termed by the authors the "Mapping situation"), while another group learned the task with all the cues clustered above the goal (the "Guidance situation"). Thus, in the first case, reaching the inconspicuous goal implied that its location had been inferred from the configuration of the cues, whereas in the second case, a visually-guided learning process was required. Only those rats from the

Mapping group were able to perform DR tests with delays of up to 30 minutes. This result demonstrates that rats from this group collected relevant information (and perhaps chose the path to be taken at the choice point) from each starting place, and were able to retain this information in memory during the delay.

Another means of better understanding the underlying mechanisms of navigation in the cross maze test consists of examining the strategies displayed at the starting places and at the central choice point. That can be done by observing behavior (latencies, places where head movements occur, duration of hesitations, and so on) in addition to subjecting animals to various transfer conditions.

In this fashion, Poucet (1985) examined the behavior of cats trained in a cross-maze according to the procedure used previously by O'Keefe and Conway (1980). The four arms of the maze were identical; three of them served as starting points and the reward was placed at the end of the fourth so that it could not be perceived from any other point in the apparatus. The maze was placed in a curtained environment and the apparatus was rotated each day in order to prevent animals from using uncontrolled extra-apparatus cues. The only available information for locating the goal was a salient visual cue. For one group, this cue was located 135° from the goal (Group 1, termed by the author the "Mapping situation"). Cats had to establish the relationship between the visual cue and the inconspicuous goal in order to get to it (Fig. 3.8, top). They could not use motor strategies since they were released from different starting points which varied pseudo-randomly. For the other group, the cue was just above the goal (Group 2, the "Guidance situation"; Fig. 3.8, bottom). A distinctive feature of Poucet's approach is that cats' behavior was carefully examined by recording the latencies for leaving the starting places, the time spent at the choice-point and the direction of head movements as well.

Both groups reached the learning criterion in a non-statistically different number of testing days. The analysis of behavioral data gave evidence of two information-gathering strategies within each group: the first consisted of taking the relevant information from the starting points (Fig. 3.8, left panel); the second was to choose the correct path from the central choice-point (Fig. 3.8, right panel). After training, DR tests were conducted with delays ranging from 8 to 60 seconds.[b] The results showed that the cats given the Mapping situation were markedly less affected than those given the Guidance situation. Among the subjects failing on the first delays in their own groups were subject A in the Mapping group and subjects K and L in the Guidance group. Interestingly, the

---

[b] The delays used in the cat experiment are far shorter than those to which rats were subjected in O'Keefe and Conway's study. This choice was motivated by the fact that restraining cats in a small starting box during several minutes induced strong emotional reactions which may have interfered with cats' spatial behavior.

behavior of those animals during initial learning was characterized by re-orientations at the choice-point. Since these cats appeared unable to take in information at the starting points, they failed to perform above chance during the delay task.

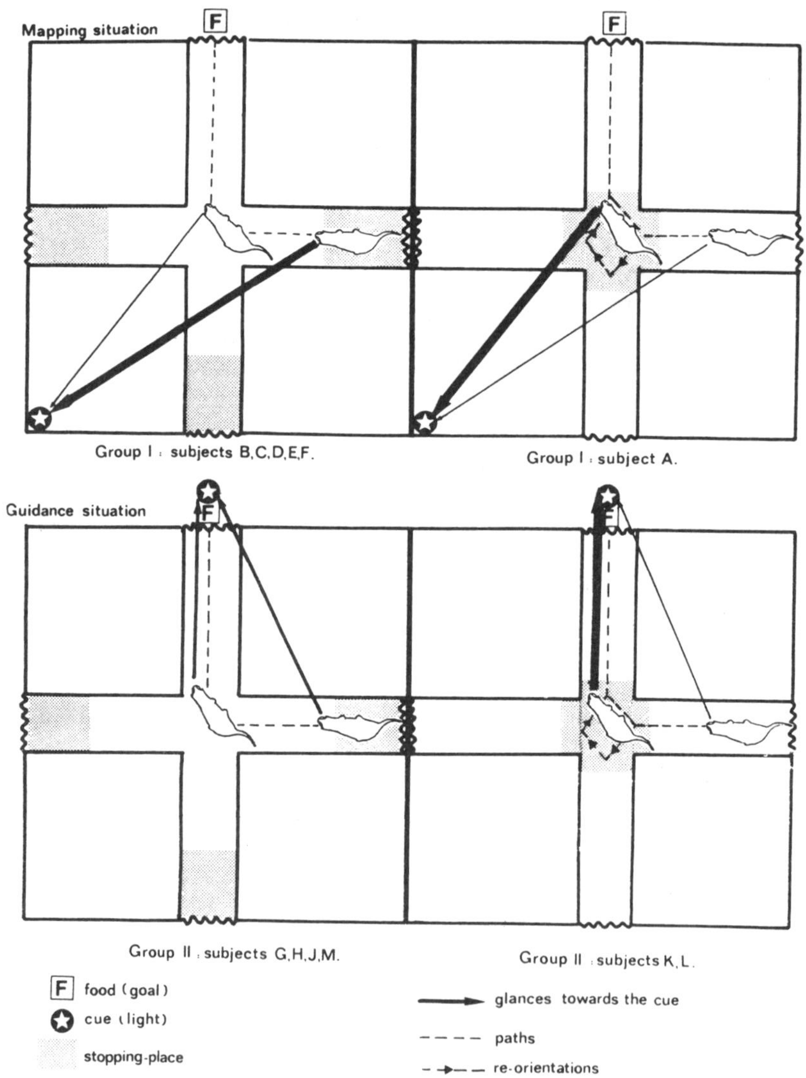

Fig. 3.8. Schematic representation of the mapping (top) and Guidance (bottom) situations with the different strategies of information gathering. The star within the circle indicates the location of the relevant visual cue. F indicates the location of the food. (Reproduced from Poucet, 1985.)

Could these animals have been successful when subjected to delays at the central choice-point where they had learned to choose their path? If so, that would mean that the very same processes are involved whatever the place where information is collected. In contrast, a failure to succeed in DR tests when rats are allowed to pick up information at the cross point would reflect the implementation of totally different mechanisms of landmark use for navigation.

In order to answer to this question, Poucet and Scotto (1987) tested hamsters (rather than cats, for practical reasons) in an apparatus similar to that used in the cat experiment. An arrangement of screens above the apparatus forced the hamsters to make their choices either at the cross-road (group II, Fig. 3.9c) or at each of the three starting boxes (group III, Fig. 3.9d). Animals from another group (group I, Fig. 3.9b) were free to make their choice anywhere as in the cat experiment. Only the situation with one unique visual cue located 135° from the goal was used.

After being trained to reach the goal from any of three starting points, the hamsters were given DR tests. The delay between the moment when they could perceive the relevant cue and the moment when they were allowed to respond ranged from 30 sec to 15 minutes. Only the performance of the subjects which could perceive the cue from the central choice point (group II) was affected by the delay. This result suggests that, although the cue was dissociated from the goal, reducing the difficulty of the task by forcing the animals to make their choice at an unique central point induced a strategy whose properties differ from those of the processes involved in the more difficult situation (choice from each starting point). The main feature of this strategy is that it does not require the animal to store relevant information for reaching the goal. In terms of O'Keefe and Nadel 's (1978) theory, failure to succeed on DR tests means that a Taxon strategy has been implemented. In the present case, it is likely that, as with some of the cats in Poucet's experiment, hamsters having to make a decision at the choice-point oriented themselves by reference to the cue (re-orientation) instead of deducing the position of the goal from that of the cue.

Thus, there is a clear correlation, or concurrence, between the strategies implemented in the localization of a goal and the resistance of performance to delays between the time that the subjects are allowed to perceive the relevant cue(s) and the moment the response is made. However, this concurrence does not provide any explanation in itself, all the less since the information which is held during the delay remains unknown. Do animals remember the location of the cue, or that of the goal, or the path to be followed? Do they compute the location of the goal as per the Vector Sum hypothesis and then keep in memory the direction to be taken? Further experiments are required to answer these questions; however, it has to be said that the label "Mapping" applied by O'Keefe and Conway, and by Poucet, to the situation in which the goal is dissociated from the cue(s) remains, at this stage, fairly hypothetical.

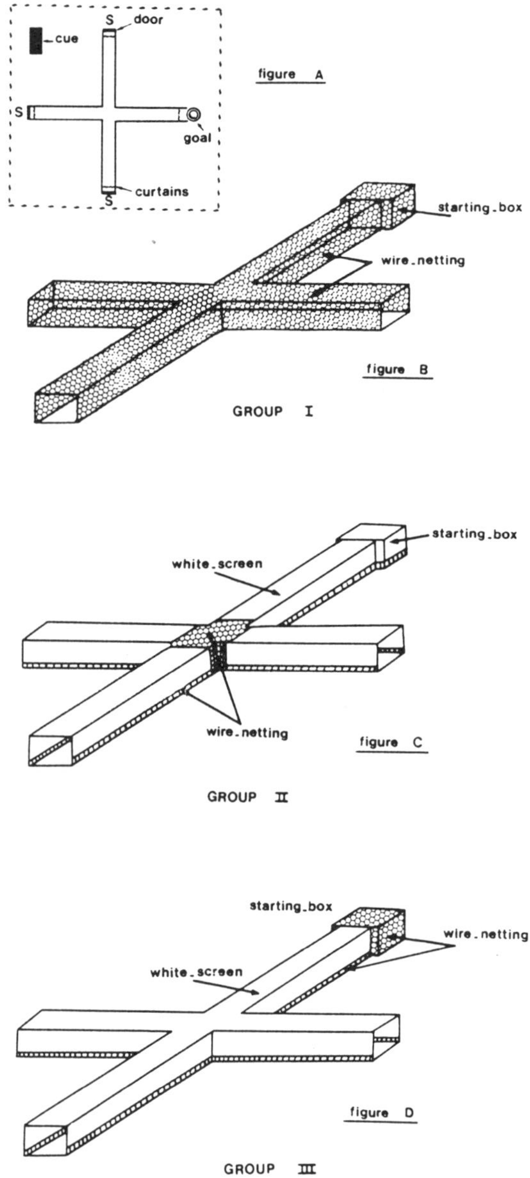

Fig. 3.9. (a) Diagram of the experimental area with the relative locations of the cue, the goal, and the starting points. (b), (c), (d) devices used for groups I, II and III, respectively. The starting box was removable so that it could be placed at several points. (Reproduced from Poucet and Scotto, 1987.)

7. The Morris Water Maze

In this place-navigation task devised by Morris (1981), rats have to escape from cold water by swimming to a small invisible platform in a large circular pool. The platform is submerged and the water is made opaque. The rats are trained to navigate to the invisible platform from one starting point only (Fig. 3.10). Transfer tests are then performed by releasing rats from other places in the pool. As early as the first transfer trial, they reach the platform almost directly. According to Morris, this instantaneous transfer to starting locations never used during training provides an example of the flexibility of spatial maps. Another condition with a visible platform is often used as a comparison; the two situations correspond respectively to what have been called Mapping and Guidance situations in the cross maze.

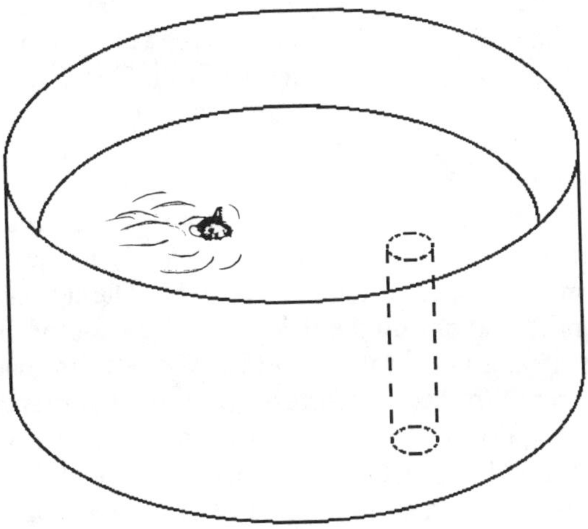

Fig. 3.10. The Morris swimming pool. The tank is filled with opaque water. The platform is below the water level so that it cannot be seen.

Accurate localization of the hidden platform involves the use of distal visual cues as demonstrated in various studies (e.g., Rudy, Stadler-Morris, and Albert, 1987; Sutherland and Dick, 1984; Sutherland and Linggard, 1982) supporting Morris' interpretation of water maze navigation in terms of mapping (see Schenk, 1996, for further discussion).

However, Sutherland, Chew, Baker, and Linggard (1987) have questioned the fact that the chosen pathway to the goal was really new, since on the first few trials, while swimming more or less randomly into all sectors of the pool, the rats had viewed the room from nearly all possible vantage points in the pool and had approached the platform from numerous directions. Thus the inference of new trajectories based upon computations from information contained within some overall map could not be assumed, and the possibility that animals use simpler strategies could not be discarded:

"*A more plausible explanation would be that the generation of a relatively novel trajectory is based upon an interpolation within a familiar range of views of distal cues and trajectory outcomes that the rat has actually experienced*" (Sutherland et al., 1987, p. 49).

This hypothesis appears to refer to a sophisticated form of re-orientation strategy – sophisticated, since visual scenes rather than one single cue would be associated with reaching the goal. In order to test their hypothesis, in a first experiment, Sutherland et al. used partitions preventing physical and/or visual access to the other half of the pool. Some groups were given a forced separate swim in the half which did not contain the platform. Thus, rats had experienced the whole apparatus but in some groups, they could not associate the perception of visual distal cues, while swimming, with reaching the goal. In the subsequent transfer trials, without partitions, rats that had been prevented from swimming in the whole pool in a continuous trajectory, or from simultaneously viewing both halves of the room, were unable to navigate efficiently to the invisible platform when released from the restricted half of the pool. The authors concluded that there are "important limitations on the flexibility of the use of distal visual information in spatial mapping when rats learn to navigate from place to place in an unfamiliar environment" (p. 56). However, given the importance of sensory and motor exploratory experience in the setting up of spatial relationships, the limited ability of rats with restricted experience of the pool may be due more to a lack of such experience than to some kind of association between a particular view and the outcome of a trajectory (see Save et al., 1996; Chapter 2). Recently, Matthews, White, Brush, and Best (1995) have found that if the partitions are faded after the rats have learned the task, instead of being removed as in Sutherland et al.'s experiment, performance is not impaired when rats are started from a novel start location. Then, some kind of exploratory and/or emotional reactions to the severe and sudden alteration to the situation may have impaired the animals during the test in Sutherland et al.'s experiment.

In a second experiment, Sutherland et al. (1987) attempted to identify the portion of the swim trial during which distal visual cues are available. For that purpose, the overhead light was switched off at the very beginning of the trial, or while the rat was swimming to the platform, or alternatively when it was on the platform. The hypothesis was as follows: if animals had learned associations between "a familiar range of views of distal cues" and the proximity of the escape platform while swimming randomly into all regions of the pool at the beginning of the experiment, preventing them from using distal visual cues either at the very beginning of the swim trial or when they had reached the platform should not have deleterious effects on acquisition. The results confirmed the hypothesis, showing that the groups for which the room was illuminated during the middle segment had shorter latencies than the other groups. Lighting conditions during the other segments had no effect.

In the same vein, Arolfo, Nerad, Schenk, and Burès (1994) studied rats' behavior in the Morris water maze using a computerized infrared tracking system which switched off the room lights when the subject entered the inner circular area of the pool where the escape platform was located. This procedure implied that rats had to swim to the platform in complete darkness after they had had the opportunity to pick up information while swimming in the peripheral ring. These authors found that rats trained under light-dark conditions found the platform more slowly than rats trained in permanent light.

Save (submitted), using a similar infrared system, found that the width of the ring, and consequently the time during which visual information was available, represented a crucial factor. After preliminary training under continuous illumination, rats were released from a starting point while the room light was on. For one group ("View group"), the light area corresponded to a narrow 15 cm-wide ring. In that case, animals were only exposed to a view of the landmark array before darkness. For the other group ("View+Loc group"), the light area consisted of a large 30 cm-wided ring. The animals from this group were allowed to swim in this area with the light on and could combine visual and kinesthetic information before darkness (Fig. 3.11). In addition, the light was turned on when the rats reached the platform. Thus, they were allowed to form "Snapshot Memories" of the target view. The data (Fig. 3.12) show that the animals subjected to the situation with the narrow light ring (View group") had longer escape latencies than the other group in both situations (View+Loc light and light/dark phase). It is striking to note that the animals which reached the goal with minimal latency times under continuous light were those mostly impaired by the short viewing duration. In addition, whatever the width of the ring, rats could see the environment once they had reached the escape platform. That did not appear to have a beneficial effect on learning. This result is in agreement with data obtained by Devan, Blank, and Petri (1992) who demonstrated that rats trained with light only while swimming to the platform reached the same level of performance as those trained in permanent light, unlike animals which had access to the visual environment only from the platform.

Taken together, these experiments all tend to demonstrate that the availability of visual environmental information is crucial neither at the very beginning of the trajectory nor when the rat is on the platform. Rather, animals appear to rely heavily on information (immediate visual feedback) collected en route, while swimming at least a part of the trajectory.[c]

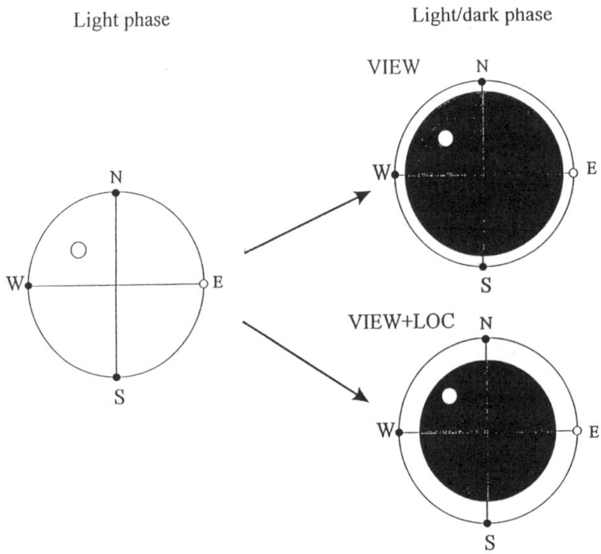

Fig. 3.11. Experimental design. Each diagram depresents a view from above of the water maze. During acquisition, rats were trained to reach the platform located in the NW quadrant from 3 starting places (N, S, and W) in the standard conditions, i.e., with the lights on (light phase) and in a condition involving alternation of light and darkness (light/dark phase). In this latter phase, rats from the LOC+VIEW group were allowed to swim on some distance from the starting points before the onset of darkness whereas rats from the VIEW group were only allowed to get a visual snapshot of the environment before the onset of darkness. (Reproduced from Save, submitted.)

These data, obtained from studies in which the light is switched off, stand in contrast with those on DR tests in the cross-maze. Indeed, both types of test require the animal to pick up information at the beginning of the trial and to keep it in memory either to control the trajectory while moving (locomoting or

---

[c] The critical length appears to be approximately 25 cm since Save did not observe any deficit for a 30 cm width of the illuminated ring, unlike Arolfo et al. with a 20 cm width. The viewing time of 1 sec used by Sutherland et al. was not sufficient to allow rats to swim a 30 cm distance.

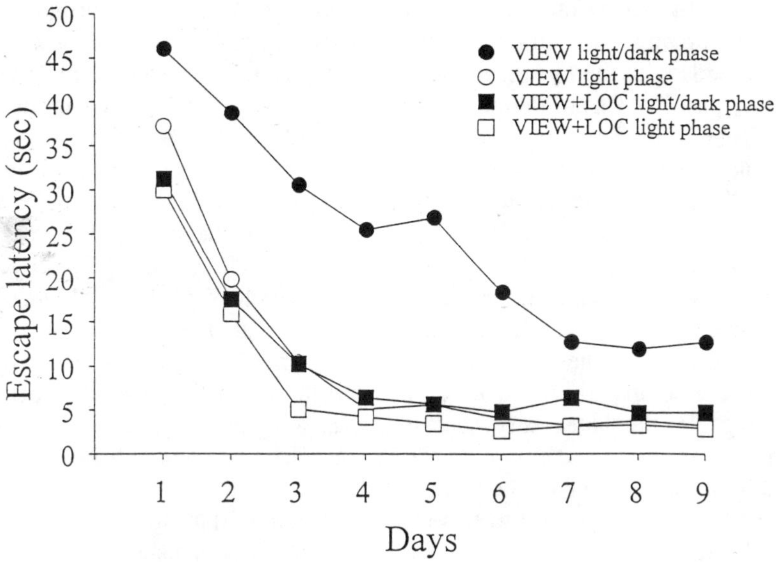

Fig. 3.12. Mean escape latency to reach the platform in seconds during training (days 1-9). The performance of each group during the light (white squares and circles) and the light/dark phase (dark squares and circles) is represented. (Reproduced from Save, submitted.)

swimming) or to plan the direction to be taken. In accordance with the conclusions drawn from DR tests in the cross-maze, we have to conclude, as did Sutherland et al. (1987), that rats rely on Taxon (re-orientation or guidance) strategies to navigate in the water maze. However, there is a major difference between the cross-maze and the water maze; in the former, runways reduce the amount of freedom in behavior and consequently the probability of missing the goal because of a slight deviation from the correct trajectory, as may be the case in the latter apparatus. We have routinely observed that rats in the water maze which have taken the correct direction without hesitation, if they miss the platform by just a little, sometimes behave as if they were lost and start searching more or less randomly. By contrast, the cross-maze task is far simpler since once the correct arm is chosen, it leads necessarily to the goal without any adjustments having to be made to the trajectory.

Another conclusion from the experiments discussed above is that viewing the environment from the platform is insufficient for its further localization when the animal is at the starting location. Such a result allows us to discard the possibility that Snapshot Memory from the goal has been used to navigate in the water maze. However, Sutherland and Linggard (1982), and later Sutherland

and Dick (1984), Keith and McVety (1988), and Whishaw (1991) demonstrated that placing the rat on the platform and allowing it to view the configuration of extramaze landmarks facilitates subsequent navigation to this location. That seems at odds with the hypothesis that the memory of the visual scene (Snapshot Memory) perceived from the platform could account for navigation in the water maze. Indeed, the between-group comparisons which have been made in each of these studies may allow us to moderate the divergence among data. When compared with animals without any preliminary experience of the maze, those which could view the environment from the platform before beginning the experiment proper, learned to reach the goal more quickly. This advantage gained by latent learning of the environmental layout is not contradictory with the fact that information while swimming is more important than that picked up from the platform. In such cases, the performance of animals not having had a view from the platform are compared with that of rats with the room light switched off while swimming. In the latter case, darkness has a much more deleterious effect which, however is not contradictory with the fact that some advantage may be gained by latent learning of the layout while on the platform.

Finally, as a conclusion to this section, it is important to stress that several types of strategy may be implemented by one animal. Whishaw and Mittelman (1986) made a careful analysis of rats' trajectories during training and subsequent probe trials with the platform removed or displaced. Their main finding is that rats can simultaneously make use of a number of strategies that are considerably more complex than had been suspected. Even rats trained with a visible platform ("cue task") used mixed strategies:

"*That they used distal cues is confirmed by the observation that on the probe trial, like rats trained on the place task, they swam directly to the platform's previous location, ... with the only cues available to guide this search being distal cues. It is also confirmed by the observation that if the platform was present, but visible in a new location, they, like place-task rats, also swim first to the old location and only then oriented and swam to the new location. That they did respond to the platform as a cue is confirmed first by their lower latencies than place-task rats, and second by the finding that they locate a repositioned platform far more quickly than place-task rats.*" (Whishaw and Mittelman, 1986, p. 428-429.)

Another unexpected finding from this study is that both cue and place-task rats returned to start points and reswam routes to the platform's previous location on probe trials (Fig. 3.13). Although the authors interpret this behavior in terms of exploration when a familiar situation is changed, it may also reflect the search for specific cues associated with the platform from the perspectives corresponding to starts, as proposed by Sutherland et al. (1987).

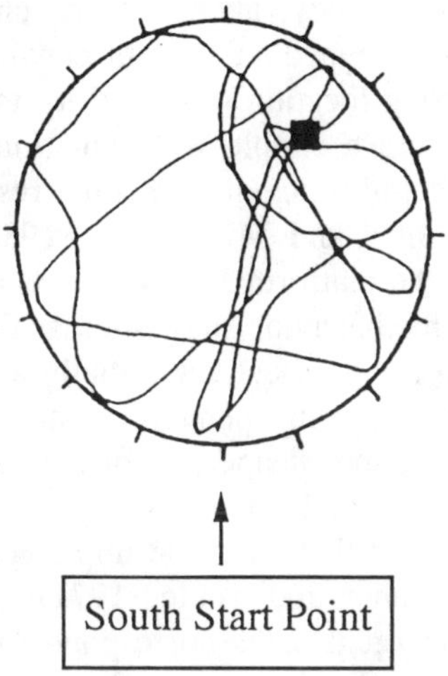

Fig. 3.13. Example of a swim pattern displayed by a rat started from 16 different start points during training, and tested with the platform removed. The previous platform location is indicated by a black square. (Adapted from Whishaw and Mittelman, 1986.)

8. Concluding Remarks

This overview of the literature reveals that a multiplicity of strategies are implemented by animals to cope with the difficulty of spatial tasks. These correspond to an optimal exploitation of the available information. Strategies appear to rely on two types of mechanism: The simplest type consists of reducing the cognitive demands of the task by using simple and economical means to process spatial information. This is the case in the multiple solution problems, when animals implement the maximum divergence principle. Other strategies, in contrast, seem to involve a higher cognitive effort that, once achieved, result in a decrease in task difficulty. Illustrative examples are provided by chunking experiments and DR tests.

The former category of mechanisms seems to run against the "cognitive map" hypothesis, or "overall spatial representations" as they have been described in Chapter 1, whereas the second one brings weighty arguments to the existence of high level spatial processing. Let us first consider those behaviors that suggest that simple means are used for orienting.

A first example is that of cats' strategies in the cross-maze (Poucet, 1985). In the Mapping group, one animal was found to choose the direction to the goal at the central choice-point. Though this non-conforming animal is not representative of the group, it is worth mentioning that its behavior had the effect of decreasing the difficulty arising from having to begin at various starting places, by implementing one unique strategy, always at the choice point. This animal *transformed* a task, initially designed to test its capability to reach an inconspicuous goal from any starting location (a capability that is taken as reflecting the use of a spatial map), into a simple re-orientation task. Though efficient during learning, this less effortful processing was not resistant to DR testing.

In the same vein, Whishaw and Mittlemann (1986) observed that in the Morris swimming pool with the platform removed, rats adopted specific patterns of swimming (Fig. 3.13). This behavior suggests that the rats attempted to perceive the relevant cues from the perspectives initially associated with the starts, i.e., that reaching the goal was not independent of starting locations. It must be said that this situation without the escape platform is a special case, and may have forced the animals to try several orientation "hypotheses" (see Chapter 1).

Another example of task facilitation is the implementation of the maximum divergence principle (e.g., Buhot and Teule, 1971). In the maze with three divergent arms, the most divergent paths correspond to the easiest locations to differentiate. Then, starting the choice sequence by visiting them guarantees that the first two runs will be correct and that food will be obtained twice consecutively, at least.

Finally, in line with O'Keefe and Nadel's theory (1978), places are not linked together by particular movements or routes but rather by various truly spatial relationships. One implication of this property is that the shape (circuitous or straight) of the possible pathways that link two places should not have any influence on spatial processing. If one defines a maze – e.g., the multiple solution problems – as a set of places having a particular status (the goals and choice point) interconnected by pathways, the configuration of the latter and the general arrangement of the places should not interfere with performance provided they do not modify the rules for solving the task. In fact, several data show that geometrical and metrical properties of the maze such as the shape of the arms (Poucet et al., 1991), the regularity of spacing between baited places (Grobéty and Schenk, 1992), the distance between them (Buhot and Teule, 1971; Ammassari-Teule and Durup, 1982; Buhot and Poucet, 1987) or the size of the central choice area (Yoerg and Kamil, 1982) can have some influence on performance levels.

Besides these examples that argue against the use of high level spatial representations, there are many others which demonstrate that animals process spatial information in such a way that their behavior becomes endowed with properties that characterize the use of spatial representations.

First, there is the consistent observation by Olton and collaborators, and also by many other researchers who have used the radial arm maze, that rats use neither stereotyped patterns of visits nor non-spatial strategies such as always entering the alley adjacent to the one just exited. That demonstrates that the goals are actually processed spatially and that animals' patterns of visits do not depend from motor strategies. This spatial processing aimed at identifying the status (baited or empty) of the goals is achieved by using a wide array of environmental cues (Mazmanian and Roberts, 1983; Suzuki et al., 1980).

Second, chunking experiments provide evidence that rats are capable of spatial categorization, for instance on the basis of the nature of the reward (Dallal and Meck, 1990; Ilersich et al., 1988). Chunking is of particular interest in the understanding of spatial representations. This capability provides an argument for the use of a high level organization of the stored information and of associated behaviors (serial order of visits), facilitating the processing and retrieval of information.

Finally, successful performance on DR tests demonstrates that when a complex spatial computation takes place (deducing the location of the goal from that of the landmark, from several starting locations), behavior is no longer under the control of permanently available relevant information, which has been stored, when it was present, for subsequent use. This independence endows spatial orientation with flexibility and adaptive properties. This would be valuable in natural conditions, for example, if the visibility of distant important landmarks happens to be momentarily obscured while an individual is navigating toward a desired place. It constitutes the basis of anticipatory and planning processes. Planning varies as a function of the structure of the apparatus and whether or not it is necessary to steer a permanent trajectory. For instance, in the Morris swimming pool, animals need to swim a certain distance from the start before they can efficiently plan the remaining part of their trajectory to the platform, whereas viewing the surroundings for a few seconds is sufficient for DR test success in the cross maze.

In summary, evidence showing that rats adopt economical and minimally effortful spatial processing is scant in comparison with the host of studies that demonstrate the opposite. However, the fact that animals are able to use an array of landmarks, to plan a trajectory, or to organize the spatial layout in various subregions are conditions that are necessary but not sufficient to prove the existence of cognitive maps. These computations can be made along the lines of Collett et al.'s (1986) Vector Sum model. The radial arm maze or the cross-maze appear likely to favour the implementation of such strategies because there is one unique choice point from which the same local views are available. In contrast, situations where the rat is released from a a genuinely novel starting location represent more powerful tests of spatial maps. A computation in terms of vectors becomes more difficult to achieve: before deducing the goal location, the animal has to orient itself by reference to the environment (see Chapter 5). Indeed, the best way to discard the possibility that a goal has been located on the basis of a local view in terms of Vector Sum computation is to design situations in which the starting location and the goal do not share the same landmarks. For instance the radial maze could be transformed into an acid test of spatial mapping if the central platform was visually enclosed, with landmarks, thus requiring that the perceived space is matched to what is beyond the perceptual field (Chapter 10).

Finally, it appears from the above data, that similar learning rates, latencies and performance levels can result from quite different underlying mechanisms. Similarly, failure to succeed may be attributed to various non-optimal strategies the implementation of which may, however, be revealing with regard to the cognitive repertoire of the species under study. In this respect, it appears essential to analyze not only traditional parameters such as the number of correct responses or latencies, but also behavioral strategies, i.e., how animals reach a given performance level. For instance, the number of errors in the radial arm maze is not very meaningful without an analysis of the patterns of visits. Latencies in the Morris task may correspond to diversified means of reaching the escape platform. Indeed, the study of swimming patterns has been found to be extremely useful in the interpretation of the effects of brain lesions (Chapter 7).

CHAPTER 4

THE PLASTICITY OF SPATIAL REPRESENTATIONS: SHORTCUTS AND DETOURS

"To the extent that we admit the possibility of representations interacting with each other, the organism is freed from the constraints resulting from a view that an association between elements is simply the summation or chaining together of those elements. The notion that representations may interact provides a basis whereby the kind of organization resulting from the interaction is more than simply the summation of the individual representations themselves." (Ellen, 1987, p. 22)

The mere ability to negotiate a detour, i.e., to deviate from a direct path toward a goal as soon as an obstacle is perceived, was considered by Köhler (1927) as indicative of "intelligent behaviors". Some species, such as hens are unable to head directly toward and around the end of an obstacle that separates them from food, persistently attempting to pass "through" the obstacle. Other studies that have been carried out to demonstrate similar effects have examined the animals' ability to take advantage of favorable alternatives; for example, the availability of a new pathway which is shorter than the usual one, allowing the animal to get to a goal more quickly by taking a shortcut.

In the preceding chapters, several examples have been given of detour and shortcut behaviors. For instance, Tolman and Honzik's 1930 experiment (Fig. 1.8) is an elegant example of detour ability in rats, while the pioneering investigations by Shepard (1931, 1933) illustrate shortcuts (see Chapter 2, and also Maier, 1932). Tests of object permanence also address such detour ability (Thinus-Blanc and Scardigli, 1981; Thinus-Blanc, Poucet, and Chapuis, 1982).

Detour and shortcut behaviors are usually taken as providing evidence of spatial representations because their achievement relies on properties some of which characterize high level spatial representations or maps: flexibility, independence of place encoding by reference to the pathways linking them, and so on. In the present chapter, an overview is given of past detour and shortcut studies. Moreover, as in Chapter 3, the data are discussed keeping in mind the following questions: To what extent do these studies bring support to the "cognitive map" hypothesis? What properties of spatial representations do they yield? And, are there explanations of the data that are more parsimonious than those based on the use of maps?

1. Making Detours: Experimental Studies

Detour abilities have been demonstrated in several species of mammals and birds (see Chapuis, 1987, 1988; Krushinskii, 1970, for reviews). The ability appears to occur as a more-or-less sudden emergence, an "insight". Whether or not detours are adopted depends very much upon the features of the experimental layout.

In this respect detour situations in which two available paths lead to the goal represent a valuable means of examining the mechanisms underlying this behavior. The angular deviation (i.e., the angle formed by the path at the starting point and the straightline direction of the goal) and the length of the paths are both factors influencing the optimalization of detour behavior. Blancheteau and Le Lorec (1972) dissociated these two parameters in the following situation: the shorter route formed a larger angle with the straightline direction of the goal (indicated by a light placed just above it) than the longer one. During early trials, the rats preferentially chose the path which formed the smaller angle with the direction of the goal, despite the fact that it was longer, but they quickly learned to take the shorter path instead.

We have used situations designed on the same principle, in which the shape and placement of barriers allowed modifications to be made both to the length of the available paths, and to their "angular deviation" from a straight-ahead path (Fig. 4.1). Moreover, a barrier could be transparent or opaque. In situation I, the two paths differ with regard to both the length and angular deviation, the shorter path making the smaller angle. In situation II, the two paths differ in angular deviation, their length being similar. The reverse is the case in situation III. Finally, situation IV represents a conflict situation, as used in Blancheteau and Le Lorec's experiment; the shorter path makes the greater angle with the direction of the goal, the longer path making the smaller angle. The tests were preceded by a phase of forced exploration of the two paths and of the food behind the barrier. Dogs (Chapuis, Thinus-Blanc, and Poucet, 1983) and cats (Poucet, Thinus-Blanc, and Chapuis, 1983b) have been given these tests either outdoors for dogs, or indoors with cats. The sizes of the barriers were adapted to the species under study.

In situation I, dogs preferred the shorter and less deviant path whether the barrier was transparent or opaque. In situation II, they displayed a preference for the less divergent path only when the goal could be seen from the starting point. In situation III, subjects preferred the shortest path when they could not see the goal. However, the most interesting results are those obtained in the conflict situation (IV). When the goal was hidden, the animals displayed a significant preference for the shorter path which also was the most divergent from the straight-line direction of the goal. However, this preference disappeared with the transparent barrier, despite animals being able to perceive the extremities of the screen that defined the different lengths of the paths. A rather similar pattern of result was obtained in cats.

It should be emphasized that the trials with opaque and transparent barriers were pseudo-ramdomly interspersed. Thus, the same animal that displayed the optimal response with the barrier occluding a part of the relevant information, was unable, a few seconds later, to choose the shorter path leading to the goal with all information available. The same phenomenon has been observed in fish by Schiller (1949). Minnows were found to better master the roundabout pathway if the goal was not visible during the approach. Young chicks also need longer time to negotiate barriers when they do not conceal the goal (Regolin, Vallortigara, and Zanforlin, 1994). Comparable data have been found by Lockman (1984, 1987) in infants (between 8 and 12 months of age). Most of the subjects executed a reaching detour around an opaque barrier before being able to do so with a transparent one.

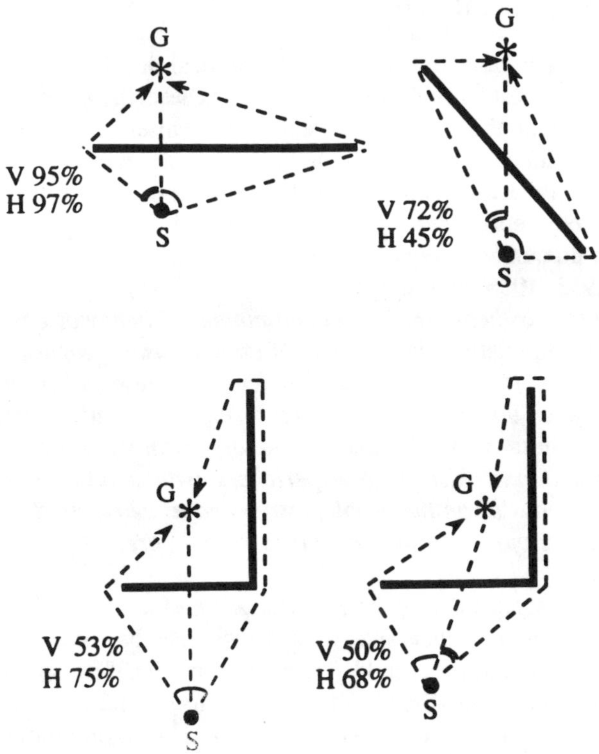

Fig. 4.1. Top view of the apparatus used in the detour experiment with dogs, and percentages of choices (adapted from Chapuis et al., 1983). S: subject; G: goal.

"*The difficulty that infants experienced with the transparent barrier was particularly striking, since, just a few seconds before, some were successfully reaching around an opaque barrier. When confronted with the transparent barrier, infants would often try to reach through it several times and then give up. Sometimes, they would even hold on to the edges and shake the barrier as if to move it out of the way. But even the tactile information regarding the presence of the edge was not sufficient to ensure success on this task.*" (Lockman, 1987, p. 252.)

These convergent results, obtained with various species, emphasize the predominance of perceived target attractiveness over the use of representations in infants and animals when they are first confronted with detour situations. Subjects have to learn to "inhibit" the attraction exerted by the target in order to implement the optimal strategy.

2. Theoretical Analysis of Detours

The detour test has been the subject of theoretical developments by various authors from the earliest period of psychology. These theoretical approaches have been reviewed in the light of current data (Amsel and Rashotte, 1984; Rashotte, 1987). Holders of both behaviorist and neo-behaviorist (or cognitivist) views had provided their own explanation about detour behavior.

One of the first theorists who gave an account of detour problem solutions was K. Lewin, a precursor of cognitivism. Rashotte (1987) summarizes Lewin's "field theory" (1933, 1935) as follows:

"*Acccording to this theory the main determinants of behavior are psychological forces that operate within the confines of an organism's cognitive representation of the problem. The way a detour problem is represented in the cognitive structures in any given instance is determined by the organism's interpretation of the problem which, in turn, is influenced by the organism's current internal state and its accumulated past history of experiences with similar problems. The organism's actual behavior on the problem is viewed as being strictly determined by the array of forces operating on the entities in the representation.*" (Rashotte, 1987, p. 41.)

Fig. 4.2a shows schematically the representation of a naive child who sees a piece of chocolate on the other side of a transparent barrier. The chocolate exerts an attractive force on the child ("force-vector" V). For this naive child, the vector m representing the direction to be taken to get the chocolate will correspond to a movement away from the desired goal. This interpretation is related to the fact that for the naive child, the experimental situation is split into two subspaces corresponding to the zones in front of, and behind, the barrier, respectively. After several attempts to get directly to the chocolate, the barrier would acquire a negative valence, favouring "insight", i.e., the sudden re-organization of a cognitive representation of the two subspaces into an unitary field. This process would allow the rearrangement of forces within the reorganized representation (Fig. 4.2b). As underlined by Rashotte (1987), this theory raises several basic questions related to the ways in which representations are constructed, to the quantification of vectors, and so on.

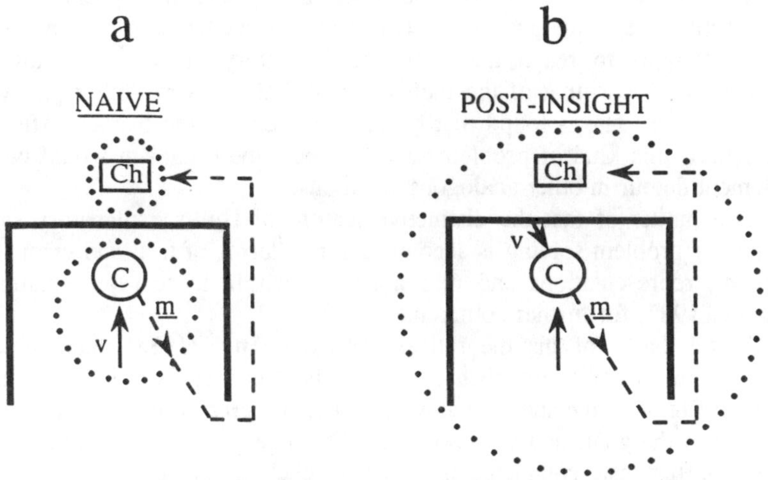

Fig. 4.2. Cognitive representation of the detour problem by a naïve child (a) and by a child who has solved the problem (b). The arrow marked V shows the direction of the force responsible for locomotion in the child's representation of the problem. C: child; Ch: chocolate. (Based on Lewin, 1933, 1935.)

Tolman's analysis (1932 a and b) of the detour problem was not radically different from that of Lewin. The degree of attractiveness of the objects (Lewin's "force vector") is called "demand-value". Some of the possible pathways to the goal are also ascribed a "demand-value". During the very first confrontations with the situation, the direct, blocked routes have the higher "demand-values". But after several unsuccessful attempts, they lose their high value to the benefit of indirect pathways. In spite of terminological differences, Lewin's and Tolman's approaches share many common features. In both theories, at the most general level,

"... *behavior in the problem situation is viewed as being determined by processes that operate within each organism's cognitive representation of the problem. Experience can alter both the content of the representation and the valences of entities in the representation, and such alterations will affect the configuration of determining processes. An insight-like process is important in restructuring the representation in a way that allows the problem to be solved.*" (Rashotte, 1987, p. 45.)

At first sight, Hull's theory (1943, 1952) does not appear to contrast strongly with Tolman's and Lewin's conceptions. The term "excitatory values of stimulus conditions" used in Hull's conception replaces those of "force-vectors" and "demand-values" of the other authors; but it is the same phenomenon, i.e., the attractiveness of the goal which is referred to in each case. Following unsuccessful attempts to reach the goal, the excitatory value of the direct path becomes inferior to that of the indirect path. Then, several other pathways are tried, until the route is found that bypasses the end of the barrier. After having experienced this kind of problem several times, the organism would be able to implement detour in other analogous situations.

As a matter of fact the distinctive feature of Hull's explanatory system is that detour problem solving is accounted for in terms of trial and error, without involving representations and reorganization within these representations (see Rashotte, 1987, for further comments).

On the bases of his theoretical construct, Hull (1938) made predictions about detour in naive organisms. For instance, concerning the situation represented in Fig. 4.3, the shorter pathway makes the greater angle from the straight direction of the goal, and vice-versa for the longer one. This situation is similar to the conflict tests conducted in rats (Blancheteau and Le Lorec, 1972), cats (Poucet et al., 1983b) and dogs (Chapuis et al., 1983), that were described above. According to Hull's predictions, a naive organism would choose the longer pathway; the longer path would have greater importance (excitatory value) because it makes a smaller angle with the direction of the goal. This is exactly what happened in rats, cats and dogs with the goal visible. Thus, Hull's prediction was confirmed by experimental data, in this particular situation.

However, when the goal is hidden, a quite different pattern of results is recorded; since the shorter path is chosen even if it makes the greater deviation with the straight-ahead direction of the goal. In Hullian terms, the excitatory value of both pathways should be the same since neither can the goal be perceived nor its location represented. Therefore, no preference should be observed. This is not what happened in the conflict situation. The recorded behavior does reflect an estimate of the goal position behind the opaque barrier. Even though they both rely on vector forces and demand-values – two notions very close to Hull's excitatory value – Lewin's and Tolman's explanations match the data recorded in our hidden goal situations; they both seem to involve representations.

As a matter of fact, whether the desired place is perceived or not represents quite distinct situations. Either a guidance is implemented (when the goal is visible) which prevents the subjects from using strategies more adapted in terms of displacement length; or a representation of the space behind the barrier may be used (when the goal is hidden) allowing choice of the shorter path. Concealing the goal releases behavior from the constraint exerted by the direct perception of the attractive object.

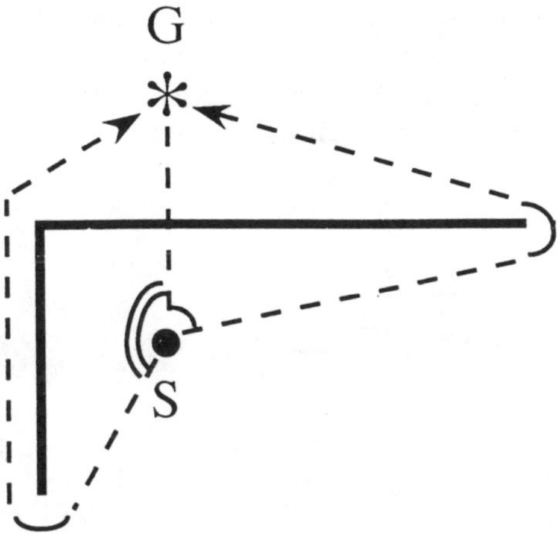

Fig. 4.3. Hull's analysis of the detour problem in which the barrier is asymmetrical. S: subject; G: goal. (Adapted from Hull, 1938.)

In summary, the ability of a particular individual to make detours is modulated by various parameters which are, primarily: whether or not the goal can be perceived, also its attractiveness value (which may correlate with the subject's state of motivation), its distance with regard to the subject, the level of familiarity of the space beyond the barrier, the presence or absence of remote landmarks, and so on.

3. Shortcut Ability

An original example of shortcut behavior is provided by Zanforlin and Poli (1970). Rats were first trained to run between two boxes, A and B, containing food and water. The boxes were linked by a translucent tubular indirect runway (Fig. 4.4). In a second phase, the rats were prevented from entering the runway, but they had the possibility of digging the sand covering the floor of the apparatus. Provided that they had been given the visual experience of the environmental layout (but not of the other goal box) from the starting place (A or B), most of them dug tunnels that converged on the goal. Following several probe trials, the authors concluded that kinesthetic information (mostly vestibular) must play a fundamental role provided that it could be "plotted" on a mental map of the environmental layout. It is regrettable that this approach has not been developed further inasmuch as it has potentially interesting properties. It represents a test of planning, since during the digging the rat has no feedback about the results of its action (an "open loop condition"). Moreover, to clear a way by digging takes longer than walking. Therefore, the required planning should extend over a relatively long time. These characteristics strengthen the significance of the results, namely the accuracy with which the tunnels were oriented.

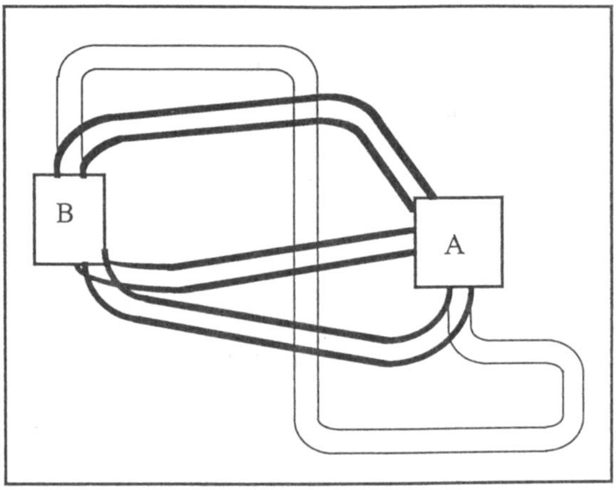

Fig. 4.4. Schemas of the tunnels dug by rats between A and B after they had experienced the transparent tubular runway (thin lines). (Adapted from Zanforlin and Poli, 1970.)

Other experiments relying on different procedures show that animals are able to display innovative behaviors provided that they have been previously allowed to explore the situation sufficiently. In an experiment with hamsters (Chapuis, Durup, and Thinus-Blanc, 1987), two subspaces, each comprising two platforms connected by an elevated runway, were separately explored (see Fig. 4.5a). Pieces of food were placed on each of the platforms. In a second phase, half of the subjects explored the pathway AB (Fig. 4.5b) connecting the two subspaces. All the animals were finally given a shortcut test for two trials. They were shown the food at D and then placed at C (Fig. 4.5c). They had three possibilities: either they could take the longer familiar pathway CBAD or two new shorter ones, CAD or CD. In comparison with the control group which did not undergo the second exploratory phase and significantly more often chose the longer familiar path, the experimental group displayed a preference for the unfamiliar shortcut CD (Fig. 4.6).

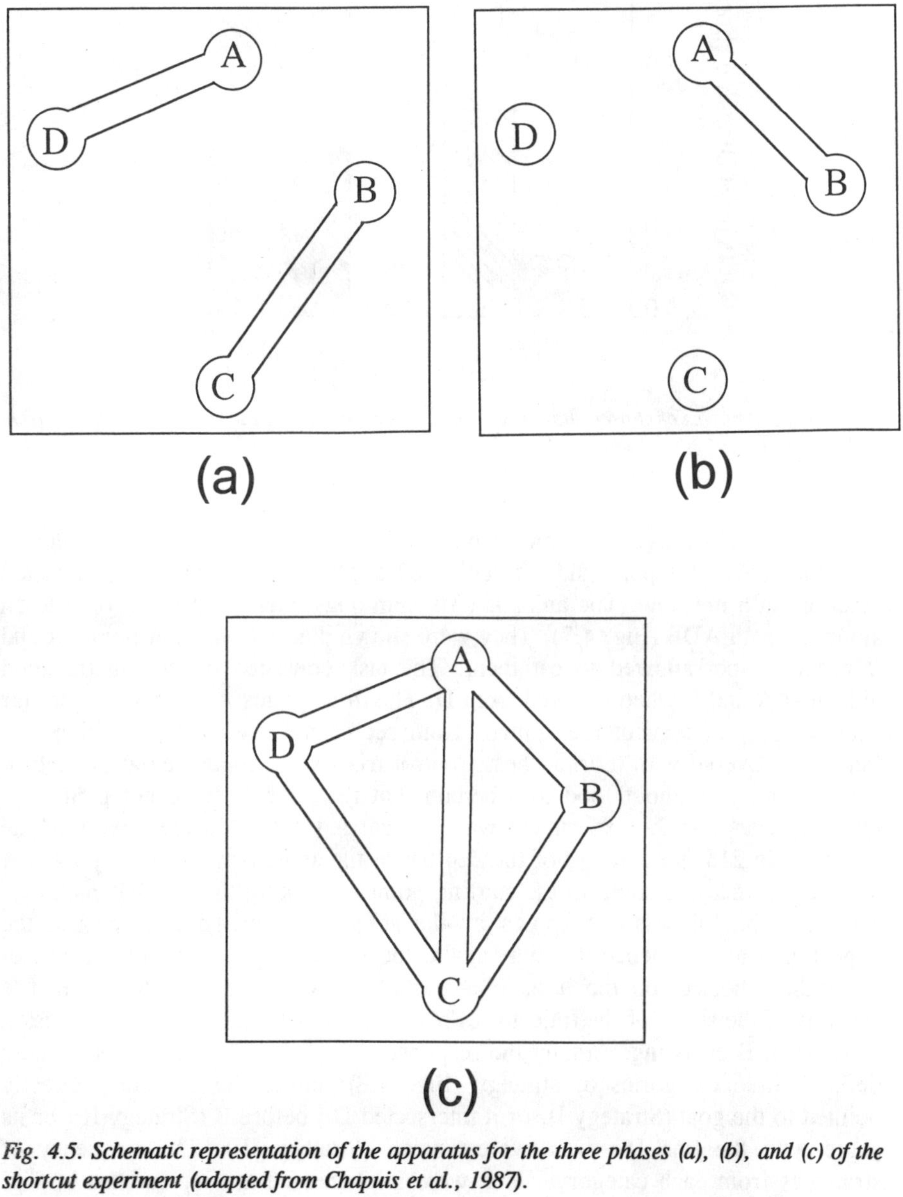

Fig. 4.5. Schematic representation of the apparatus for the three phases (a), (b), and (c) of the shortcut experiment (adapted from Chapuis et al., 1987).

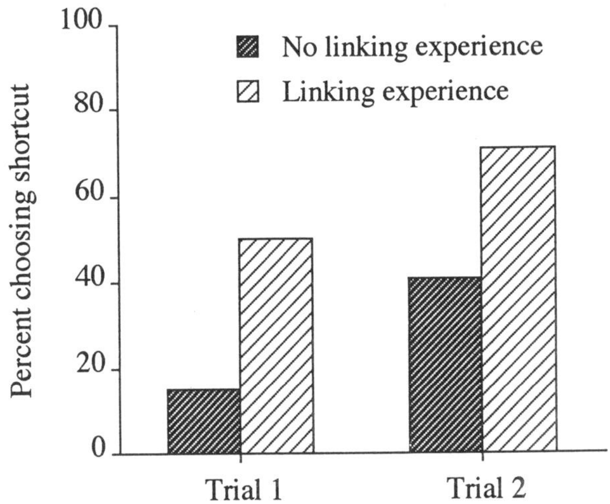

Fig. 4.6. Percentages of chosen shortcuts in the two groups during the test trials (adapted from Chapuis et al., 1987).

Similar innovative capacities have been demonstrated in dogs in another shortcut study (Chapuis and Varlet, 1987). The experiment was conducted outdoors in a meadow. The animals (Alsacian dogs) were first taken on a leash along the path ADB (Fig. 4.7). They were shown pieces of meat at points A and B but were not allowed to eat them. The task consisted of finding the food hidden at A and B when released from D. The three points A, B, and D were far enough apart to prevent the animals from seeing the reward. In addition, the field was covered with thornbushes. Control trials were conducted at the end of the experiment without food to ascertain that the animals were not guided by olfactory cues and the experiment was also carried out in different parts of the meadow. In 215 (96 percent) of the 224 trials, the animals went directly from A to B rather than returning to the starting point and taking the familiar pathway. The segment AB was not known by the subjects which have never had the opportunity to experience it. Thus, again, the animals were able to work out an unfamiliar shortcut on the basis of a partial knowledge of the situation. The analysis of the shape of the trajectories revealed, however, a relative imprecision of location B encoding. Among the responses labelled as "correct", the authors defined three categories of strategy (Fig. 4.8): either the trajectory directly pointed to the goal (Strategy I), or it intersected DB before B (Strategy II), or its angle with the line DA was overestimated (Strategy III). The frequency of strategies from each category which were implemented are represented in Fig.

4.8. Interestingly, there were only a few trajectories corresponding to Strategy III, which would lead the dog into an unfamiliar region.

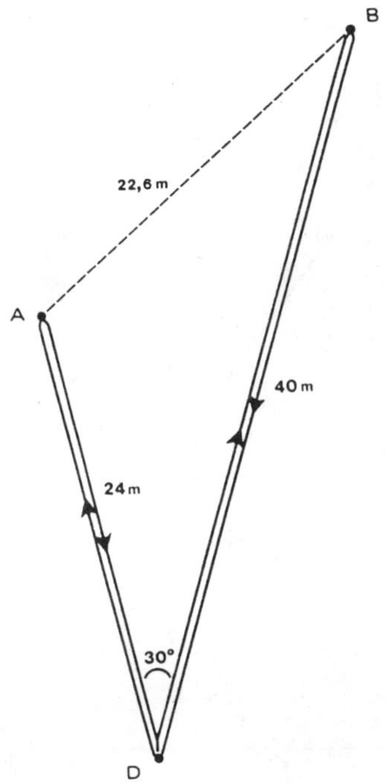

Fig. 4.7. Schemas of the situation used with dogs. D: starting point. A and B: food points. DA-AD-DB-BD vectors (solid): exploratory runs (dogs on leash); AB segment (dotted): optimally oriented task run. (Reproduced from Chapuis and Varlet, 1987.)

Shortcut tests were conducted by Poucet in his 1985 experiment with cats (see Chapter 3). Alleys directly linking the four extremities of the maze were added and shortcut abilities were tested from starting places S1 and S3 (Fig. 4.9). Whether they belonged to the Mapping or Guidance group, only cats that had been found to choose their path from the starting point during learning succeeded in taking shortcuts; subjects that chose from the central choice-point needed to pass through this place when a shorter path was available and seemed to ignore the new possibility of reaching the food faster by adopting a more economical route.

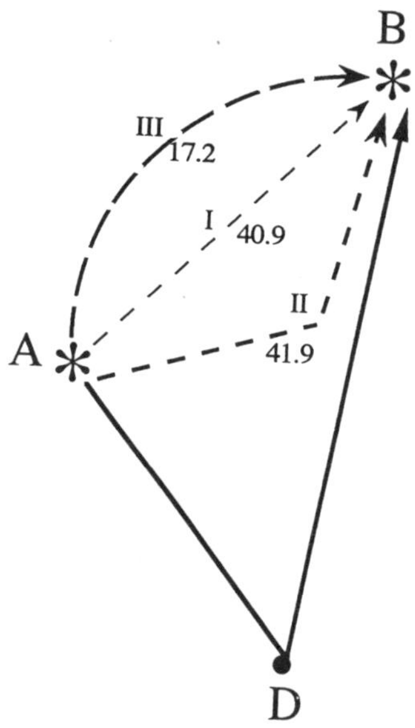

Fig. 4.8. Illustration of the different categories of dogs' shortcuts and frequencies of choices. (Adapted form Chapuis and Varlet, 1987.)

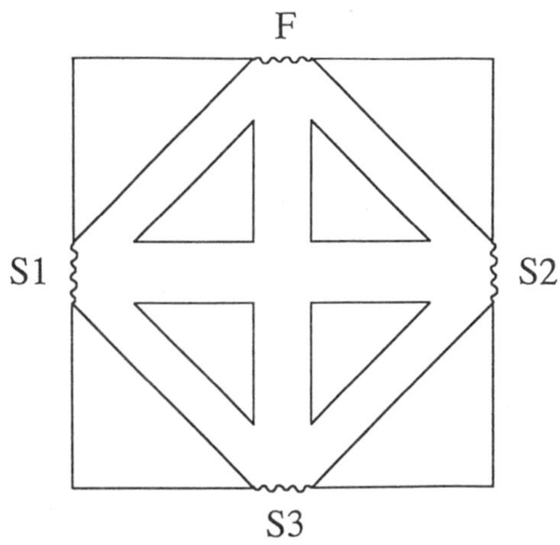

Fig. 4.9. Overview of the modified apparatus used in the shortcut experiment. (Adapted from Poucet, 1985.)

4. Concluding Remarks

What are the mechanisms underlying detour and shortcut behaviors? What do these behaviors demonstrate about the properties of spatial representations?

It has been established that shortcuts and detours show oriented behavior to be relatively independent of a given pathway. However, this plasticity does not have the same meaning in shortcut and detours situations, and also varies according to whether the goal or associated landmarks are visible or not. When the goal can be seen, detours are possible but they are not optimal if two paths of different lengths are available. The behavior remains tightly dependent on the degree of attraction exerted by seeing the reward. In contrast, provided that animals have not learned previously to make their choice at a specific place other than the starting point, as in Poucet's experiment (see above), seeing the goal facilitates the choice of the shortcut that, in this specific case, relies on guidance. It consists purely of the subject minimizing the distance between itself and a desired object. Indeed, with a visible goal, motivational aspects obscure the true spatial ability of the subject to make a shortcut or a detour.

In contrast, that animals can make successful shortcuts and detours when the goal is hidden is a strong argument for the idea that they have mastered the notion of equivalence of pathways. Once spatial relationships are established between places and charted on a representation, there is more than one pathway, like that which is usually taken, which can link them together. The orientation process does not rely on rigid routes but on the precise localization of the places concerned. It follows from this that if the possibility of a shorter path is offered, pointing in the direction of the goal, it is chosen without hesitation. Similarly, if a detour is necessary, the subject is able to monitor its position while deviating from the straight direction of the goal. In the latter case, in addition, deviating from the straightline goal direction in the first part of the trajectory implies an anticipation of the outcome of this action. It is necessary that the animal disregards its immediate motivation to get the food as directly as possible in order that the best choice of route can be made.

Finally, another property, more obvious in shortcut than in detour behaviors, is that spatial information can be re-organized since it does not depend on the sequence in which it was acquired. That is particularly true in the case of shortcuts with a hidden or distant goal as in Chapuis and Varlet's experiment. The observed behavior does not rely on a route. That implies in turn that information is accessed in a simultaneous form and is independent of the sequence in which acquisition took place. Other examples of inferential processes such as the early studies by Maier (1932; see Chapter 2) show that animals are able to "break" a learned sequence of places provided that they have been allowed sufficient exploration.

In summary, all of the above mentioned properties – plasticity, equivalence of pathways, anticipation and inferential processes – are those that characterize cognitive maps as they have been defined by Tolman and by O'Keefe and Nadel (see Chapter 1). However, as in the traditional learning situations, one question remains, that of the nature of the landmarks that are used to solve the task. Needless to say, if a conspicuous landmark is located just behind a hidden goal, guidance strategies may be implemented and none of the above properties is relevant. In many past experiments, the environmental cues have not been controlled. However, in Chapuis and Varlet's study, for instance, the experiment was conducted each day in a different location in the test ground. Thus, each day, a different distant landmark would have become associated with the goal at the point when the dogs were shown the food. Given the short duration of the tests, this does not appear very likely to have occurred. In the detour experiments with cats and dogs, the experiment took place each day at the same place, but trials with opaque and transparent barriers and with different arrangements of the barriers and of the starting and goal locations were intermingled. Again, the possibility that animals used guidance strategies is rather unlikely.

Another possibility is that animals could have computed the direction to the goal as per the Vector Sum model (see Chapters 3 and 5). If the same distant landmarks can be perceived from both the starting place and the goal, the possibility cannot be discounted that the location of the latter is inferred on the basis of stored vectors. The only situation where animals had no access to distant visual landmarks while orienting is that used in Zanforlin and Poli's experiment where rats dug in the ground to reach the goal by a straighter pathway than that experienced initially.

Data from the experiment in hamsters (Chapuis et al., 1987) presented above argue against the implementation of a guidance or a vector sum strategy. The control group, which did not explore the linking path, did not preferentially take the shorcut at the very first test. If animals had used distant landmarks common to the starting place and the goal, all of them should have taken the shortcut immediately. Indeed, those animals from the group which had explored the linking path chose the shortcut more frequently than the controls.

These observations suggest that guidance or vector sum strategies were probably not implemented by animals tested in detour and shortcut situations described above. The other alternative is that high level spatial representations or cognitive maps have been used by animals to solve the tasks.

CHAPTER 5

PSYCHOLOGICAL MODELS OF SPATIAL PROCESSING

"One basic problem with cognitive-map concepts is that the survey-map characteristics they posit in memory have no obvious connection to the traveler's actual perceptual experience of the environment... That is, the survey-map is a bird's eye view of the environment, whereas what we experience is a frontal view." (Strelow, 1985, p. 241)

In parallel with the empirical study of orientation and map construction, using studies such as those described in the foregoing chapters, theoretical models have also been developed. Models are theoretical constructs which attempt to incorporate and fit together as many pieces (experimental data) of the jigsaw puzzle (spatial cognition) as possible. In my opinion, they should not be regarded as describing the process itself but rather what it could be. Even if they do not always explicitly lead to experimentally testable predictions, they have an invaluable heuristic function.

I have arbitrarily chosen to discuss in the present chapter some of the models which take into account animals' actual experience. It is a matter of fact that an environment is perceived in the form of many frontal or local views that are egocentrically organized. Some of these percepts are stored in memory and used for orientation. This process cannot be denied. In contrast, spatial representations, whether they are map-like or not, are the psychological result of the integration of animal's actual perceptions. We can only deduce from the observation of spatial behaviors the presumed nature of underlying representations. For these reasons, I consider that the models of spatial representations which put emphasis on the subject's actual experience are likely to be more psychologically plausible than those which ignore that level.

The "Vector Sum" model already alluded to in Chapter 3 is the simplest since it selectively concerns inferential processes and computations based on a perceived visual scene. Vector computations involve a very precise evaluation of distances, and they are likely to be successfully achieved when the animal is close to both the relevant landmark(s) and the goal. However, in natural conditions, animals often have to travel much longer distances. In those cases, other means of orienting are necessary. Some authors have proposed simple mechanisms for navigation within relatively large spaces, that may share some basic features with O'Keefe and Nadel's (1978) routes but they do not amount to the same thing. Two of these proposals are summarized in the present chapter: McNaughton's Conditional Association model (1987, 1989) and Lieblich and Arbib's World Graph model (1982). Section 4 is devoted to "dead-reckoning" which is a central notion in Gallistel's (1990) theoretical developments discussed in the next section. Finally, Poucet's (1993) model of spatial processing is presented.

1. The Vector Models and Landmark Use

Many examples of experimental situations with landmarks that are widely used for orientation have been provided in the preceding chapters. In the case when an inconspicuous goal is very close to a conspicuous landmark, guidance strategies are implemented. Such strategies consist of minimizing the distance between the landmark and the observer. However, in other situations, the desired place is not closely associated with a particular landmark. Computing the direct and optimal trajectory requires that spatial relationships between one or several perceived landmark(s) and the non-perceived goal is known. For instance, in Chapter 3, I have discussed some data which show that rats use landmark configurations in the radial maze to differentiate among baited and un-baited arms (e.g., Suzuki et al., 1980). However, these studies do not tell us anything about the way these landmarks are used.

Collett et al. (1986) have examined this issue in gerbils by using a method which consists of modifying the array of landmarks. These experiments and the related model provide a possible explanation of how landmarks may be used to deduce goal location.

In Collett et al.'s (1986) experiments, gerbils were trained to find buried food reward (sunflower seeds), the position of which was specified by an array of one or several landmarks in an otherwise visually featureless arena. Once training was complete, tests were conducted without seeds and the landmark array could be modified. The place where the animals searched for the food was recorded. Collett et al. have demonstrated that gerbils are able to search in the correct location when learning had taken place with seeds placed at a constant direction and distance from a single cylindrical landmark. But the fact that gerbils still succeeded when released from various starting places suggests that they had used cues other than the cylindrical landmark. Since the latter was identical when approached from any direction, external cues were necessary to provide the bearing of the goal with regard to the landmark, otherwise searching would have occurred within an annulus around the landmark. By modifying the size of the landmark, the authors also showed that goal localization is independent of the landmark's perceived size but relies on actual distances. Some of the most compelling data supporting a precise distance encoding is that gerbils can still reach the goal defined by an array of landmarks when the room light is extinguished during their approach. This means that the animals had already planned the length of the trajectory to the goal before the room was darkened (which in turn implies that distances between the landmarks and the goal had been precisely stored in memory).

Collett et al. have put forward the hypothesis that vectors exist which are necessary to localize a desired place before planning a route. This proposal deserves special attention since it represents one of the few attempts to define the inferential processes involved in the localization of an inconspicuous goal. Two vectors are involved in this calculation: a "seen" vector corresponding to the animal's position with respect to the landmark and a "stored" vector which gives the position of the goal with respect to that landmark. Thus a direct trajectory to the goal can be computed by the gerbil, by calculating the difference between the two vectors (Fig. 5.1, top). According to Collett et al., only one landmark is necessary for the correct localization to be computed. However, for this computation to be accurate on the basis of a single landmark, a general "compass" direction is necessary when the animal approaches the landmarks from various directions.

On the basis of the demonstration that gerbils are able to separately process various landmarks, Collett et al. propose another means by which the animals could check that a correct trajectory has been computed. When two landmarks are used, they are processed separately and two similar trajectories result from two distinct computations. If both the vantage point from which the landmarks are perceived and the stored vectors are in correspondence, each computation should indicate the same goal location. If not, they will indicate different goal locations. Consider the landmark configuration in Fig. 5.1 (bottom). If the two trajectories computed separately from A and B and based on the stored vectors "the goal is located 5 m south of A" and "the goal is located 6 m north of B" point to the same place, then the animal knows that this is the right direction. If the computation results in two different goals being defined, this means that it cannot be correct.

In addition to this astute way of verifying that the computation is correct, Collett et al. consider three possibilities likely to provide the animal with directional information: (1) the goal specified by one or several landmark(s) is always approached from a constant starting direction; (2) "compass" cues from distant visual landmarks or from various non-visual cues specify the orientation of the layout when the animal approaches the goal from various routes; and (3) directional information can be provided by the geometrical configuration of the array itself, when there are several landmarks. An additional simpler possibility, not raised by Collett et al., is that landmarks have an obviously different appearance according to the perspective from which they are perceived.

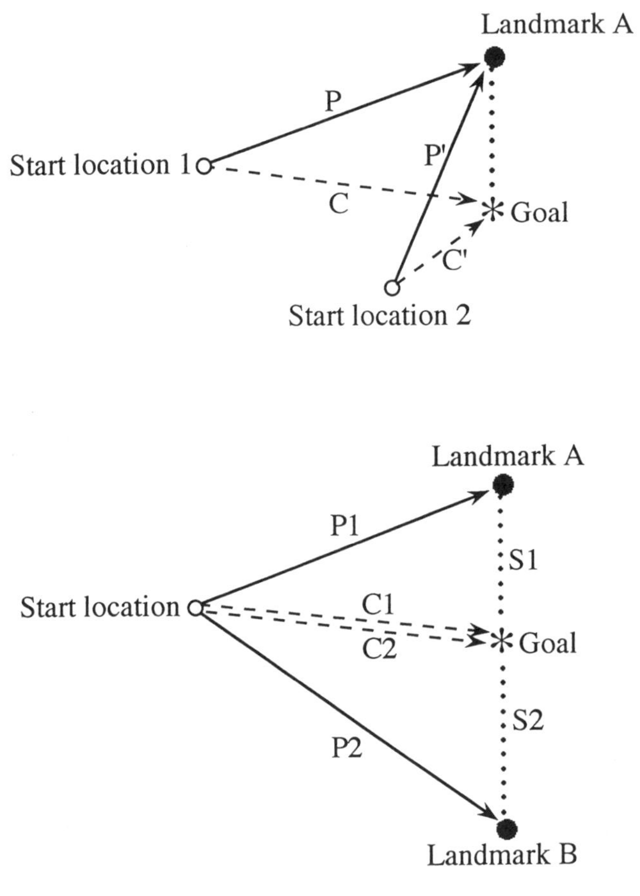

Fig. 5.1. The use of vectors to plan a trajectory. Top: the trajectory (C) to the goal is specified by the difference between the seen (P) and stored (S) vectors associated with one landmark (A) when the subject is at Start Location 1. The same type of computation takes place when the subject is at Start Location 2. Bottom: when there are two (A and B) landmarks, a separate, but coincident trajectory is specified by each. (Adapted from Collett et al., 1986.)

Natural conditions usually provide an abundance of multimodal local and distant landmarks, having different appearances according to the direction of approach, and which can be perceived as forming perceptually salient geometrical configurations. It is quite possible that the various methods that an animal could use to locate itself with regard to the useful landmark(s) are implemented according to their availability.

Moreover, many data demonstrate that general compass information, common to several places and situations, is actually used. The nature of the geocentric compass appears to widely vary across species. For instance, one particular category of visual information comes from the sun. Its accurate use relies on the association between sun azimuth, time of day and geographic direction. This "ephemeris function" implies the existence of an internal clock indicating time of the day. In fact, an azimuthal position of the sun at 45° with regard to the horizon would not enable the differentiation of east and west, unless an internal clock allows the animal to distinguish between morning and afternoon. The sun compass is usually invoked to account for long distance orientation in birds. However, recently, Wiltschko and Balda (1989) demonstrated that such information can be used by food caching birds (scrub jays) in short distance orientation in an outdoor aviary. Their data strongly support the idea that the birds used sun compass orientation to relocate their cache in spite of the various available environmental visual cues which were ignored. However, in closed rooms, under artificial lighting, Balda (1987, quoted by Wiltschko and Balda, 1989) had shown that scrub jays are able to recover food at a rate above chance level. In that case, the birds must have used environmental cues. These data, taken together with others obtained in homing experiments (Fuller, Kowalski, and Wilstschko, 1983), suggest that when sun information is available, birds use it predominantly. However, as pointed out by Wiltschko and Balda (1989) if the sun compass were the sole source of information about the direction of cache sites, birds would be unable to locate them in overcast conditions. In addition, in the absence of further experimental data, one must be extremely careful about the generalization of sun compass orientation to species other than birds (see Benhamou and Poucet, 1996).

Another source of compass information could come from the geomagnetic field. Like the sun compass, this source of information has been invoked for long distance orientation in animals and in human subjects as well (Baker, 1987). However, so far, there has been no evidence of the use of this type of cue in short distance orientation. Etienne (1987) showed that after a passive outward journey from their nest to the center of an arena, hamsters were still able to orientate toward the location of the nest in spite of a tenfold reduction of the earth's magnetic field combined with the fact that other types of information (visual, acoustic, olfactory and tactile) had simultaneously been rendered irrelevant.

In summary, although the use of some kind of geocentric compass has not been demonstrated in rodents, different types of information appear likely to provide the animal with a general sense of direction, permitting them to deduce the trajectory to a goal on the basis of vector computation.

Thus, the Vector Sum model corresponds to some kind of "completion" of a spatial layout by computing the position of a hidden element (the goal) which belongs to that layout. Since this computation can be made from any direction of approach, the model strongly suggests that animals have extracted the spatial invariance of the situation, i.e., that they know the relationships between the landmark(s) and the goal independently of the vantage point from which they are perceived. That seems to correspond to the use of a spatial map which would encompass only a few elements, the goal and one or more landmarks (each of them being associated to the goal by a distinct stored vector).[a] By its very nature, a map can be used from any starting place, even one that has never been experienced (see Chapters 1 and 3). In this respect, it would be interesting to examine in Collett and al.'s situation the behavior of the animals during the very first test trials, when they are released from new starting places. Does the computation of the goal location take place immediately or is learning necessary? In the former case, an immediate transfer would indicate that a somehow abstract relationship between the goal and the subject's location has been charted on a restricted goal-centered survey map, regardless of the direction of approach. In contrast, the necessity to learn to compute a trajectory from a new vantage point would mean that goal localization relies on specific frontal views and that as many vectors have to be learned and stored in memory as there are directions of approach. This is a parsimonious explanation which would deserve to be tested experimentally.

2. Linking Places with a Minimal Cognitive Cost

McNaughton's Conditional Association model. McNaughton (1989) and McNaughton, Leonard, and Chen (1989b) have proposed that spatial relationships between successive views are encoded during exploration in terms of the simple movements that connect them. For instance, if a left turn from location A leads the animal to perceive location B, then a conditional association between making a left turn from location A and the representation of the view of location B would be established (Fig. 5.2). The animal would learn the conditional relationship between turning left when coming from A and the sensory consequence (reaching B) of that turn and thus, could represent and expect the local view resulting from its action in this particular context. According to McNaughton and Nadel (1990), the chaining together of such conditional relations between movements and views would keep the animal continuously informed of its position by reference to the last seen local view. Thus, S-R-S associations are learned. They

---

[a] Nevertheless, unlike cognitive mapping theories, such a goal-centered model of spatial memory ignores the processing of features that are not directly related to the goal location (Poucet, 1993).

would ensure a rough, though step by step, integration of sequentially perceived local views and associated movements, which may lead to the computation of equivalent pathways (Fig. 5.2).

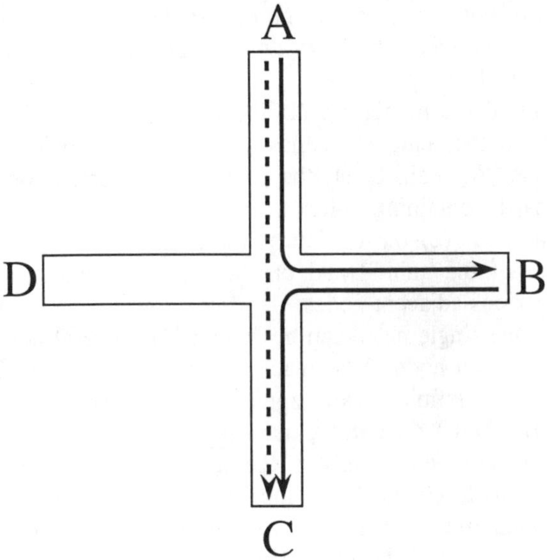

Fig. 5.2. Illustration of the implementation of the Conditional Association model in a cross-shaped maze. A, B, C, and D are distinguishable landmarks that specify locations. A and B, and B and C are linked by a left turn. On the basis of these two consecutive left turns from A, the animal could compute that they are equivalent to a straight trajectory to C. Note, however, that the model does not take into account the 180° turn that the animal has to make at B. (Reproduced from McNaughton et al., 1989b.)

This model has the advantage of assisting with the specification of the spatial functions of some brain structures (see Chapter 9). However, reaching a place (in the sense of encountering a particular local view) exclusively relies on a given direction of approach that is likely to provide the expected next local view: it does not seem to allow the flexibility that is required to enable a goal to be reached that is not itself an element of the learned chain of S-R-S associations.

Lieblich and Arbib's World Graph model. The "World Graph" (WG) model proposed by Lieblich and Arbib (1982) is aimed at providing a functional definition of spatial maps. Schematically, a WG is made up of a set of nodes corresponding to "recognizable" places. Nodes are linked by edges representing a path from one recognizable place to the next. A distinctive feature of Lieblich and Arbib's model is that:

"*... a single place in the world may be represented by more than one node in the graph – as a result, for example, of the animal coming upon a place for the second time but not "recognizing" that it has been there before, perhaps because it encounters the place in a very different situation or motivational state.*" (Lieblich and Arbib, 1982, p. 628.)

Other features of the model are (1) the weight given by the authors to the visual system and to "perceptual schemas" assembled to provide a representation of the environment. At brain level, the perceptual schemas correspond to egocentric visual maps containing information about the position of the objects within the visual field; (2) the choice of the next node to reach relies on the motivational state of the animal and on the expected reduction in motivation associated with various adjacent nodes.

The fact that one single place can be depicted by several nodes, and that the choice of such-and-such node to be reached depends on motivational state, run contrary to traditional definitions of cognitive spatial maps (see Poucet's model and Chapter 1). By their very nature, map-like representations hold information about absolute places, independently of the subject's current motivational state, and of the "perceptual schemas" (or local views) which they provide. In his comment on Lieblich and Arbib's target article in *Behavioral and Brain Sciences* (1982), Kaplan underlines that for the WG model to account for overall representations, it does require a mechanism that integrates into a single node the various views of the physical elements which characterize a place. In other words, the detection of "object constancy" is required.[b] In the absence of such an abstract form of processing, the WG appears to account for an intermediate level of spatial representations, characterized by nodes (labelled by perceptual schemas) and linked by functional paths between nodes.

Unlike McNaughton and al.'s proposal, the WG does take into account the possibility of making choices between the next node to be reached according to the motivational state of the animal but it does not take into account the necessity for a subject to update position changes while moving among nodes so that it keeps informed of its current position by reference to an environmental anchor.

3. Dead-Reckoning and Path-Integration

In 1873, in the Journal *Nature*, Darwin had advanced the hypothesis that some animals would be capable of navigating efficiently by using a mechanism

---

[b] Indeed, this basic notion should be termed in this case "place constancy".

that was well known to sailors in that epoch to determine the position of their ship from the record of the course sailed. This would be especially important in the absence of any other source of information, for example from the sky or from a coastline. Provided that an overall direction was known (by using a compass), the length of the journey could be computed from its speed and duration. Combined with the changes of direction indicated by the compass, it was thus possible to keep track of the ship's trajectory. In a simplified form, this model of navigation has inspired many experimental studies about the role of "path-integration"[c] in spatial knowledge. During locomotion, two main categories of information are provided. The former category consists of stimuli (visual, auditory, etc.) coming from the external world whereas the second, that "feeds" path-integration is made up of all the forms of internal feedback that are directly generated by locomotion itself, i.e., muscular, tendinous, plantar and vestibular inputs.

Path-integration has been studied extensively in hymenopterous insects, which use the sun as a directional reference (Wehner and Wehner, 1990). In nocturnal mammals, such as gerbils (Mittelstaedt and Mittelstaedt, 1980), dead-reckoning relies entirely on self generated internal cues. But the most extensive studies of path integration have been conducted in the hamster by Ariane Etienne and her collaborators. The choice of the hamster was justified by the fact that when this animal encounters a substantial amount of food, it fills up its cheek-pouches and is strongly motivated to return directly to its nest site whether or not the outward journey had been circuitous. In Etienne et al.'s studies, each subject's nestbox was permanently connected to an arena located in a cue-controlled environment (Fig. 5.3). A heap of hazelnuts was placed at the center of the arena. The animal coming from its nest filled its cheekpouches with hazelnuts and had then to return to its nest under several conditions. It was found that in the case of conflict between route-based and location-based information, the latter prevailed provided that it was conspicuous enough (Teroni, Portenier, and Etienne, 1987). For instance, when used separately, simple landmarks or a patterned background are insufficient for visual cues to predominate. It is necessary that landmark objects are presented against the patterned background for the visual information to predominate over self generated information (Etienne, Joris, Maurer, and Teroni, 1990a).

---

[c] Path-integration and dead-reckoning rely on the same mechanisms, i.e., on the integration of movement-generated information. The former term is usually employed about short-distance trajectories in animals, whereas dead-reckoning more often refers to long-distance navigation in humans.

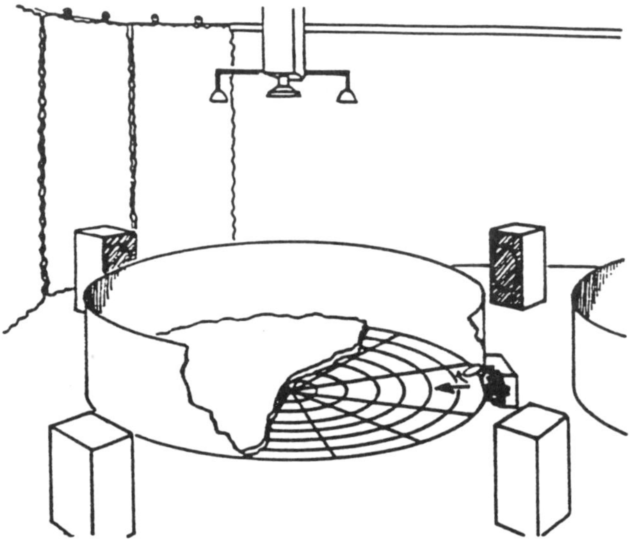

Fig. 5.3. Experimental arena used by Etienne, Maurer, and Saucy (1988). The hamster's nest is permanently connected to the arena. Marks on the floor allow the experimenter to record hamsters' trajectories. The floor is covered with sawdust which is turned over from time to time in order to prevent animals from using olfactory scents deposited during previous journeys. A camera is mounted above the arena and 4 loud-speakers emit white noise for masking possible auditory cues. (Adapted from Etienne et al., 1988.)

The moment when the visual cues are available is also an important factor; for instance a light spot exerts its influence mainly during the phase of food collection, just before the subjects start to return to their nest. However, its effect remains noticeable even when it is presented only briefly at the beginning of the hoarding excursion (Etienne, Teroni, Hurni, and Portenier, 1990b). In addition, active or passive rotations while the animal is at the food source are compensated for within certain limits (Etienne et al., 1988).

The general aim of these studies was to define the conditions under which an accurate return journey toward the home can be made on the basis of an initial outward travel to the food source. One of the interests of these results is that they demonstrate at a general level the importance of internal (idiothetic) information for spatial knowledge since idiothetic information is efficient even if the displacement cannot be perceptually inscribed within an allothetic frame of reference. As emphasized by Etienne (1987), and Poucet and Benhamou (1996) the integration of idiothetic information does not result in objective spatial relationships among places but lead to a spatial knowledge which is referred to the origin of the whole trajectory. Like all navigational systems which depend on route-based information, path-integration and dead-reckoning can generate cumulative errors, and therefore remain reliable over limited distances only, in the absence of external information that permits errors to be corrected. In natural conditions, dead-reckoning allows foragers to return safely to their home, the central point from which most of their displacements usually radiate and on which they usually converge. When external cues are available, they interact with path-dependent information to enhance the precision of spatial representations (Etienne, Joris-Lambert, Reverdin, and Teroni, 1993; Georgakopoulos and Etienne, 1994) and contribute to geocentric representation (see next section, Gallistel, 1990). During exploration linked to foraging and other functional contexts, dead-reckoning may therefore play an essential role.

4. Gallistel's Model

Gallistel (1990) has recently proposed a two-step model of mathematical computation by which egocentric representations could be transformed into geocentric maps. It is summarized in the following terms:

"*I suggest that the surveying procedures by which animals construct their cognitive map of the environment may be broken down into two inter-related sets of processes, which, together permit the construction of geocentric metric maps. One set of processes constructs a metric representation in egocentric coordinates of the relative positions of currently perceptible points, lines, and surfaces – a representation of the geometry of what is perceived from one's current vantage point. The other process, dead-reckoning, provides a representation in geocentric coordinates of the vantage points and angles of view (headings). Combining geocentric representations of vantage points and angles of view with egocentric representations of the segment of the world perceived from each vantage point yields a representation of the shape of the behaviorally relevant environment in a system of coordinates anchored to the earth, a geocentric cognitive map.*" (Gallistel, 1990, p. 106.)

The egocentric representations derive directly from what the subject has perceived and depend on the subject's current location; this notion seems to be very close, if not the same as that of "local views". Gallistel, however, places a special emphasis on the *metric* and *geometric* features of the visual layout and cites experimental evidence showing that many species are effective in perceiving the angular and linear distances of stimulus sources. Similarly, although it has received less attention than the use of discrete landmarks, the geometric shape defined by a set of objects or landmarks is a spatial feature that is spontaneously encoded by animals.

An elegant demonstration of the ability to encode geometric relationships comes from a study by Cheng and Gallistel (1984). They used an elongated X-maze placed in a rectangular enclosure. A patterned card was attached to the side in each corner of the enclosure (cf. Fig. 5.4a). The distal end of each arm of the maze was baited with a number of pellets which differed from one arm to

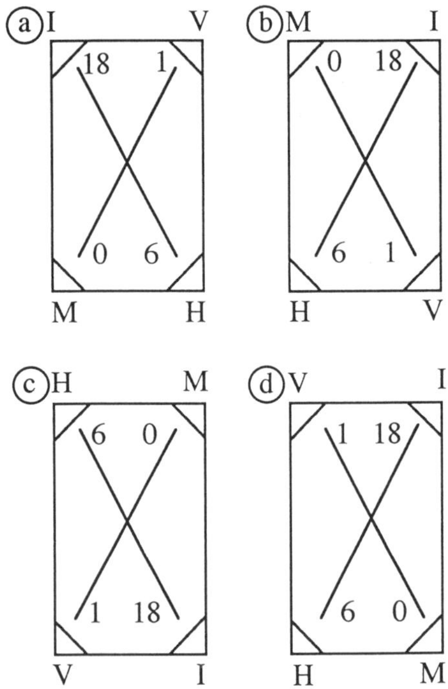

Fig. 5.4. The experimental arrangement used to test the rat's reaction to various transformations of the environment. a: the training set up. The letters written just outside the rectangle represent the landmarks, which are, in fact, mounted on the diagonals in the corners of the rectangular enclosure. The numbers beside each arm of the X maze indicate the number of pellets with which that arm is baited. b: affine transformation of the training setup. c: Euclidian transformation (rotation). d: reflection. (Reproduced from Cheng and Gallistel, 1984.)

the other (from 0 to 18 pellets). Therefore, each quantity of reward was associated with a different pattern. The rat was placed at the intersection of the arms and allowed to collect the food. The order of the visits corresponded to the amount of food contained in the arms. After this training phase, the respective locations of the four patterns were modified according to affine (Fig. 5.4b) or Euclidian (Fig. 5.4c) transformation, or the pattern arrangement was reflected (Fig. 5.4d). The amount of food associated with each of the patterns remained the same throughout these tests. Therefore, if the rats had learnt to associate an amount of food with a given pattern, their performance should not have been disrupted by any manipulation which left these associations unchanged. Instead, performance dropped drastically in conditions b and d which affected either the distances between the patterns (b) or brought about the reversal of their arrangement (d). Following the rotation which affected none of these features (c), the animals chose sequences optimally as in the control condition (a). It is to be noted that the change represented in Fig. 5.4b did not modify the relationships between the different cues. The only modification consisted of change in the distance between two cues. The fact that this modification had such a disruptive effect on performances demonstrates that the rats had encoded the geometrical relationships between patterns and that this kind of "Gestalt" was predominant over other features. With slightly different procedures and apparatuses, Cheng (1986, 1987) and Margules and Gallistel (1988) confirmed that rats rely primarily on geometrical features of the environment.

Other data bring additional support to the encoding of geometrical properties. For instance, in the dishabituation experiments (see Chapter 2), we observed a renewal of exploration in all of the situations where the geometry of the object arrangement was modified after habituation (Poucet et al., 1986; Thinus-Blanc et al., 1987). In contrast, enlarging or decreasing the size of the configuration without changing the geometrical shape did not produce any reaction. Studies conducted in the radial maze provide further examples of the incident nature of maze geometrical properties (shape of the arms: Poucet et al., 1991); regularity of spacing between baited places: Grobéty and Schenk, 1992; etc.) on performance levels (see Chapter 3). All together, these results strengthen the idea that geometrical relationships represent an important spatial feature for the animals, that may, in some cases predominate over or interact with the use of discrete landmarks.

The metric and geometric representations in egocentric coordinates of what the animal has perceived at a given moment concern discrete places ("segments of the world") that, according to Gallistel, are integrated into a global cognitive map by dead-reckoning:

"*Dead-reckoning relates the animal's position and heading during one observation to its position and heading at another, which is what makes it possible to compute a nonegocentric representation of the relative position of objects and surfaces perceived at different times from different vantage points*" (Gallistel, 1990, p. 58.)

As underlined in the preceding section, the whole process has a unique origin, the nest for instance, which is used as a geocentric reference point. All the subsequent "en cascade" computations rely on the integration and record of the travels which have been made from that origin. The principle underlying this computation is to relate the linear displacements and re-orientations to the unique origin. Hence the perceptions of the observer, which are egocentrically organized, could be integrated into a geocentric frame of reference whose metric properties have been precisely encoded while walking from one vantage point to another.[d]

Thus, according to Gallistel, an animal knowing its approximate position on his map would be able to anticipate the next image to be encountered. A process of comparison between the current image and the anticipated one leading to the best match between the two, would thereby ensure a correct step-by-step displacement. This hypothesis could be tested by charting the animals' patterns of displacements within a large scale environment. But this comparison process is not by itself a distinctive feature of maps. It can be implemented while following a route (in that case the risk of mismatch between the current and anticipated images is minor since the motor response is supposed to bring the subject to the correct vantage point where the next motor response will be produced). It properly accounts for keeping one's bearing during long-distance displacements in which relays or "reassuring cues" (Lynch, 1960) are necessary within familiar environments. In contrast, it is more difficult to understand how this type of representation could come into play in the computation of a new trajectory when the situation is partly known such as in detours and shortcuts.

Indeed, the most important feature of Gallistel's hypothesis is that maps allow the accessing of a representation of an image corresponding to a place on the map. The relationships between the level of action (perception and representation of egocentered local view) and that of overall representations built up on the basis of the former, are at the core of the model. In this respect, the proposal is appealing and realistic since it takes into account the necessary reciprocal links between the actual experience and the resulting psychological constructs which, in turn, provide information to be transformed into action for oriented behaviors to take place.

---

[d] The construction of cognitive maps as proposed by Gallistel appears to some extent rather germane to what I have called "extraction of spatial invariants" in Chapter 2: movement-generated information ensures the observer of the stability of the world and of its invariant features, in spite of its versatile appearance due to perspective changes.

5. Poucet's Model

Poucet has recently (1993) proposed a hierarchical organization of space which is based on the necessary steps that would be required for building map-like representations. First, the integration of the various local views available from one single place during eye, head and body rotational movements yields the formation of a *place representation*. This integration allows place identification and recognition independently of any specific local view, i.e., independently of any direction of approach. Among the studies dealing with this important issue, that described by Olton (1979) is of particular interest since it demonstrated that rats are able to approach a goal from a totally new location. Getting to the goal also implied moving in a direction that had never been experienced before (Fig. 5.5).

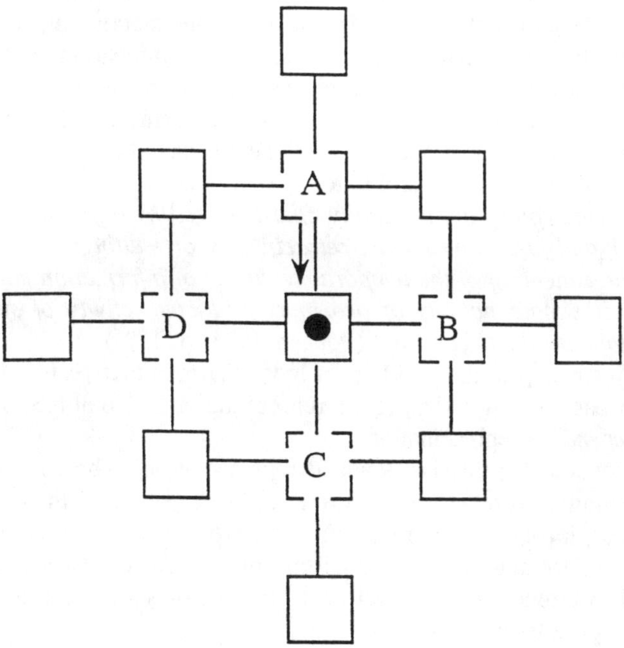

Fig. 5.5. Schema of the apparatus used to test rats' ability to reach the goal (black dot) from a totally new direction of approach. The rats were trained from the starting points D, B, and C. During the test, they were released from A. (Reproduced from Olton, 1979.)

The second step of Poucet's model consists of establishing connections between place representations by encoding vectors that hold directions and distances. Distance information can be computed on the basis of vision and movement-related parameters derived from locomotor activity. In contrast, a general direction of reference is necessary for directional information to be useful in complex spatial computations. Poucet's assumption is that instead of using one unique compass-like geocentric direction, animals compute several location-dependent reference directions which allow them to determine directions from any particular location.

In a subsequent step, place representations which share a number of common stimulus elements provided by either the proximal or distal environment constitute "local charts". In each local chart, a number of privileged locations serve as reference points from which the direction and distance of other points can be determined. These are *location-dependent* representations (Fig. 5.6). The encoding of topological relationships (proximity, connectivity, inclusion, order) is necessary for planning trajectories between connected elements. An interesting feature of this topological network is that new information can be deduced by combining existing information. For instance, a combination of two or more simple trajectories between distinct places can be replaced by a single and shorter one (shortcut), since general rules for processing space, such as associativity and transitivity (Piaget and Inhelder, 1956), have been derived from experience and, in particular from exploratory experience:

"... *because exploration is usually characterized by a pseudo-random activity (with back-and-forth movements, retracings, short-cuttings, etc.), it ultimately leads to the canceling of the temporal sequence of information gathering, with the result that simple notions of proximity and connectivity of space become readily available to the organism.*" (Poucet, 1993, p. 172.)

The ultimate step in the build up of local charts is that vectorial information contained in each location-dependent representation is combined into a global *location-independent* representation.

A pre-requisite for this last level of representation to be constructed is that there is one unique direction of reference. Poucet suggests that while moving around and exploring the environment, some places are more important than others. This differential attention to certain places would lead to the emergence of an overall reference direction allowing for vector summation by providing a common reference to all places.

This process accounts only for short-distance orientation but not for navigation between places that do not possess any stimulus elements in common. To deal with this constraint, the model assumes that local charts are interconnected. Either the goal to be reached is in a local chart that shares a "linking" element with that of the start (Fig. 5.7a) or there is a number of intermediate local charts connected one to another by two or more linking elements (Fig. 5.7b).

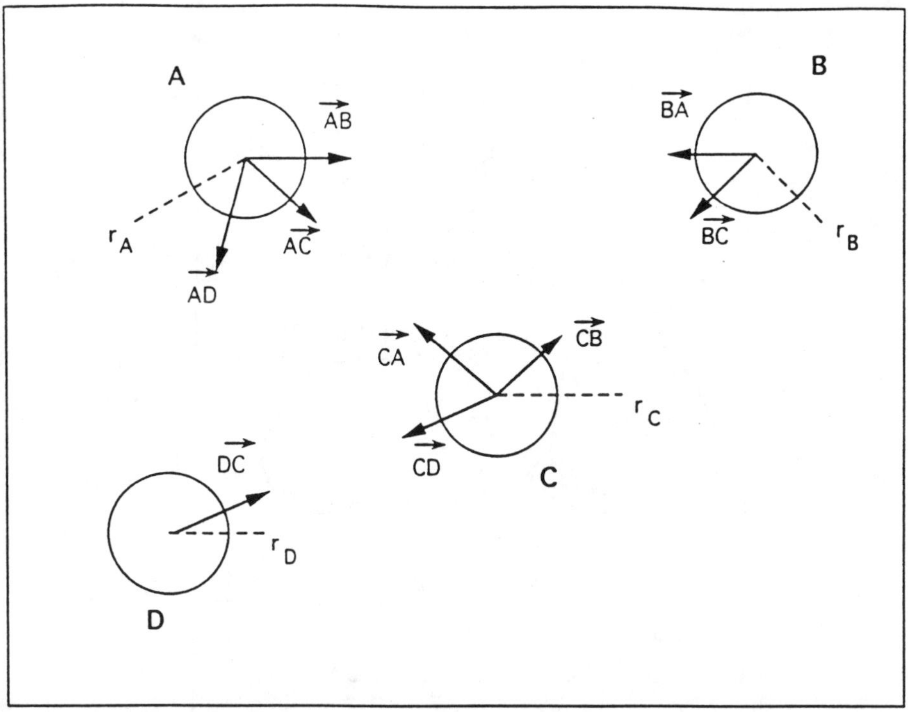

Fig. 5.6. Schematic representation of a location-dependent reference framework. Each place (indicated by A, B, C, D) provides its own reference direction (rA, rB, rC, rD). Vectorial information about distant places is therefore location-dependent. Note that the availability of vectorial information about a distant location (e.g., about location D) from another location (e.g., from location A) does not imply the availability of vectorial information about the latter location from all other distant locations (in the given example, there is no vector from D to A). Yet, some locations may provide vectorial information about all other locations (for example, as a result of their central position in space, location C). (Reproduced from Poucet, 1993.)

However, the trajectories will not necessarily be direct and will still depend on the pattern of connections between local charts. For a direct trajectory to be computed (Fig. 5.7c), it would be necessary that "... the distinct reference directions provided by different local charts be combined into a single, local chart-independent, overall direction so that a two-stage vector summation would be sufficient for correct orientation to a distant environment" (Poucet, 1993, p. 173.)

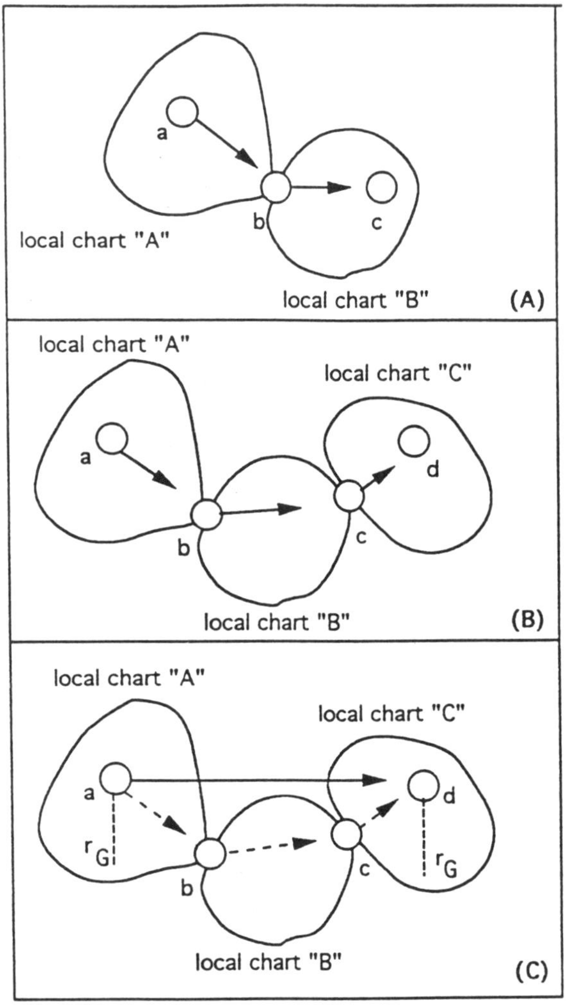

Fig. 5.7. Connections between distinct local charts. In panel A, the problem of reaching location "b" (local chart "B") from location "a" (local chart A) is solved by using local chart "A" as a location-dependent reference framework which provides information as to the linking place "b" between the two non-overlapping environments. panel B shows how such operations can be combined so as to solve spatial problems (e.g., reaching "d" from "a") between two non-overlapping environments "A" and "C" with no linking place. Note that the resulting trajectory is not direct. In contrast, the availability of a general reference direction (rG) independent from any location or environment, allows for the computation of a direct path from "a" to "d" (panel C). (Reproduced from Poucet, 1993.)

Although animals' capability for achieving this highest level of spatial processing and orientation is still controversial and has not yet received experimental support, this model has several advantages. For instance, as Poucet points out, it permits the optimization of cognitive resources because of its hierarchical organization (see Yoshino, 1991, for a demonstration of this advantage). Moreover, accurate orientation can be achieved, under specific circumstances (i.e., within local charts, among places that share common environmental stimuli or landmarks) without having recourse to map-like representations holding the metric properties of the physical space. Rather, an important weight is given to topological characteristics. That does not prevent the system from processing more abstract spatial information. Another important feature of the model is that it makes a distinction between the mechanisms involved in orientation in proximal space (within local charts), and long distance navigation, respectively. For instance, whereas general rules for performing spatial operations based on associativity and transitivity (detours and shortcuts) can be implemented within a local chart with landmarks common to both the starting place and the goal, an additional pre-requisite (a single, local chart-independent, overall direction) is necessary for steering long distance navigation across several local charts.

6. Concluding Remarks

As mentioned in the introduction, only some models which rely primarily on the representation of the environment as it is actually perceived by animals, are examined in the present chapter (see O'Keefe, 1991b, for a summary of other models). Important elements stem from these proposals which consider both the notions of place and of links between places at levels of varying complexity and abstraction.

The first element for constructing a map or, more generally, a spatial representation, is that places can be identified. In this respect, the notion of "local view" is explicitly used by some authors (Poucet, McNaughton) whereas others prefer more detailed terms: "current view of a landmark array" (Collett et al.), "egocentric representation of what is perceived from one's current vantage point" (Gallistel), or "perceptual schemas" (Lieblich and Arbib). All of these terms refer, however, to egocentered (usually visual) scenes which are tied to the subject's current position in the environment.

In many of these models, local views are not themselves the object of particular processing (McNaughton, Lieblich, and Arbib) but, rather, they are simply considered as stored "tags" for identifying places. Although Gallistel emphasizes the metric and geometric features of what is perceived from a current vantage point, the status of the resulting egocentric representation does not seem to basically differ from McNaughton et al.'s expected local views. In contrast, for

Collett et al., the "current view of a landmark array" is the substrate for the computation of the location of an inconspicuous element, the goal. Finally, it is in Poucet's model that local views are processed at a higher level, as the basic constituents for extracting the notion of place representations.

The second level of spatial representations is based on the idea that places (or that local views that label places) are connected one to another. The simplest possibilities are that local views are linked together by elemental movements (McNaughton et al.) or by paths (Lieblich and Arbib). In Gallistel's model, dead-reckoning ensures linkage between various vantage points, allowing for the computation of a geocentric map independent of the relative position of objects and surfaces perceived from various perspectives. Finally, according to Poucet, connections between place representations are vectors that have directions and distances, combined with the encoding of topological relationships.

Besides the question of a more or less abstract definition of places and of links between them, other important features of these proposals have to be underlined. One of them is the distinction put forward by Poucet between the representation of restricted parts of space, the local charts, and that of large scale environments where landmarks are not common to all places likely to be linked by trajectories.

Another issue that is considered in some models is the encoding of topological versus metrical spatial properties. Unlike some authors who incorporate the two types of encoding into their theoretical developments (e.g., Poucet), Gallistel places a special emphasis on the metric and geometric encoding of space at both levels of spatial organization, that of egocentered representations and that of geocentric maps. Indeed, angles and distances contained within a given local view can be caught by one glance. For example, the use of "snapshots" in bees as demonstrated by Cartwright and Collett (1983) presupposes a very accurate encoding of angles and distances to be efficient. Many species (Thinus-Blanc, 1975) and even nocturnal animals such as rodents (e.g., Thinus-Blanc, 1978, 1981, 1982) are skillful in discriminating distances and, a fortiori, so are higher mammals and human subjects. However, even the latter display a very poor encoding of metric properties of map-like spatial representations of large scale environments (e.g., Lynch, 1960) contrasting with the finely tuned representation of metric properties of visual images (for a review, see Denis, 1989).

The above remarks do not contradict the data demonstrating that geometric properties are spontaneously encoded and used by animals. But if goal localization during the final part of a trajectory, based for instance on vector computation (Collett et al., 1986), requires the perception and storage of both precise angle and distance relationships, a similar precision of distance estimate is not necessary and has never been demonstrated in mammals for long distance navigation. In the latter case, a general reference orientation allowing the animal to determine the direction to be taken appears to be a crucial feature, by comparison with distance estimate, at the onset of the travel.

In conclusion, a basic idea to keep in mind when attempting to understand spatial processing is that the same mechanisms cannot account for spatial behavior in both restricted and large environments because each provides distinct types of information, which may result in the adoption of distinct spatial processing mechanisms. For instance, within a small area, the same landmarks are often available from the starting point and the goal, and this clearly requires a type of processing different from that needed when two regions of space are visually "disconnected". Correspondingly, the importance of metric encoding varies as a function of the length and complexity of the trajectory to be computed. Moreover, whereas various reference directions may be used within restricted parts of space, enlarging the area or the number of subspaces where trajectories take place implies the use of a reference direction common to all the traversed subspaces.

Finally, it should be mentioned that recent years have witnessed the development of computational models which, despite the fact that they rely on data from psychological and/or neurobiological studies, emanate from mathematics, computer sciences and robotics. The discussion of these models is beyond the scope of this book. However, these mathematical accounts of spatial representations and navigation incorporate important ingredients for our better understanding of spatial mechanisms. For instance, some of them are based on the notion of local views (Schölkopf and Mallot, 1994), on distinctions between views and places (Schmajuk and Thieme, 1992), or between local and global spatial knowledge (Yoshino, 1991), or they postulate the existence of networks which consist of partial representations of the world, elaborated by multiple interacting sub-systems (Prescott, 1996). All these notions are of major interest and are likely to help psychologists in their progression even if, in my opinion, the time is not quite ripe for direct exchanges between these different disciplines.

CHAPTER 6

THE HIPPOCAMPAL FORMATION AND SPATIAL PROCESSING

> *"Recording the activity of neurons can only tell us what signal transformations the cells might be performing, however; they tell us that they are sufficient for the task, not whether they are necessary. Therefore single-unit findings must be supplemented by careful analysis of the outcome of lesions, in order to confirm that what we believe the cells might be doing is eliminated when they are destroyed." (Stein, 1991, p. 205)*

A major event in the field of spatial cognition and its neurobiological bases has been the discovery of a population of hippocampal neurons, the properties of which reflect unambiguously the encoding of the animal's location in the environment. They were christened "place-cells" by O'Keefe and Dostrovski who discovered them in 1971. Despite some debates about the precise meaning of place-cell firing properties, and also about the involvement of such cells in non-spatial processing, this finding has given a strong impetus to research in the domain of the brain substrates of spatial cognition. A few years later, O'Keefe and Nadel (1978) proposed that the hippocampus represented the neural substrate of cognitive maps. As already discussed in Chapter 1, the integration of findings about brain functioning and spatial behavior into a unified theory strengthened the impact of the initial discovery, giving rise to a growing interest in rodents' hippocampus and spatial memory. Some years earlier, the observation by Scoville and Milner (1957) of the dramatic temporally graded amnesia produced in H.M. by bilateral lesions to the hippocampus and related temporal lobe structures had a comparable general impact in the field of human memory. This chapter is aimed at giving an overview of studies which allow us to better understand the role played by the hippocampus and associated structures. Studies of the effects of lesions on traditional learning tasks, exploratory behavior and spatial problem solving tests, are presented in the first two sections. The other sections are devoted to electrophysiological studies.

1. The Hippocampus: a Structure at the Crossroads of Brain Functional Networks

The hippocampal formation includes several structures related to the hippocampus proper (Cornu Ammonis, or Ammon's horn): the dentate gyrus, the subicular complex and the fornix.[a] The hippocampus is the site of convergence of pre-processed multimodal sensory information (entering via the entorhinal cortex and perforant path) from all of the cortical mantle, and of movement-related information (via the septum, generating theta rhythm, and via the fimbria-fornix). It returns projections both to the neocortex and many subcortical structures.

Although less extensively studied, septal lesions usually induce deficits similar to those of hippocampal lesions in comparable spatial tasks (for a review, see Gray and McNaughton, 1983). For that reason, in some studies, the septum and hippocampus are together referred to as the septo-hippocampal system.

The general organization of the hippocampus has been described in many reviews (e.g., Foreman and Stevens, 1987; McNaughton and Nadel, 1990; Miller, 1991; O'Keefe and Nadel, 1978; O'Mara, 1995). The hippocampus proper (Cornu Ammonis) is subdivided into four main cytoarchitectural fields, CA1, CA2, CA3 and CA4 that are unidirectionally connected from CA4 to CA1 (fields CA2 and CA4 are small and not well defined). The organization of the main intrahippocampal connections has been classically described as a trisynaptic loop (Andersen, Bliss, and Skrede, 1971). Schematically, (1) the entorhinal cortex via the perforant path sends projections to the granular cells of the dentate gyrus, (2) which themselves project by mossy fibers to pyramidal cells of CA3 (and CA4). Finally, (3) Schaffer collaterals link CA3 to CA1 pyramidal cells. Then, axons from CA1 project to the subiculum, the entorhinal cortex and, more sparsely, to more distant cortical regions. Whereas the entorhinal cortex represents the most important route of entry for neocortical afferents to gain access to the hippocampus, the subiculum appears to be the main way of exiting for information, once it has passed through the trisynaptic loop (Fig. 6.1). The connections between the various hippocampal fields define relatively independent lamellae, conceived as functional units for information processing, aligned and roughly perpendicular to the longitudinal axis of the hippocampus (Fig. 6.1). These lamellae are related by fibres which connect them in the longitudinal direction. It should be emphasized, however, that this is an oversimplified account of hippocampal circuitry. On the basis of recent findings, O'Mara

---

[a] The terminology (hippocampal formation, hippocampal system) and the structures concerned may vary among authors (see Eichenbaum, Otto, and Cohen, 1994). The definition adopted here is that of current use in studies of rodent hippocampus.

(1995) has recently provided an up-dated picture of intra- and extra-hippocampal connections (Fig. 6.2), which does not question the existence of the trisynaptic loop (see also McNaughton and Nadel, 1990, for a discussion of this issue) but underlines the importance, among others, of direct projections from the entorhinal cortex to CA1, and of several recurrent circuits (e.g., CA3-CA3 connections):

"*It is more accurate to describe the entorhinal input and the hippocampal projections as being somewhat laminar with extensive cross-laminar projections and extensive crosstalk between the 'laminae'.*" (O'Mara, 1995, p. 254.)

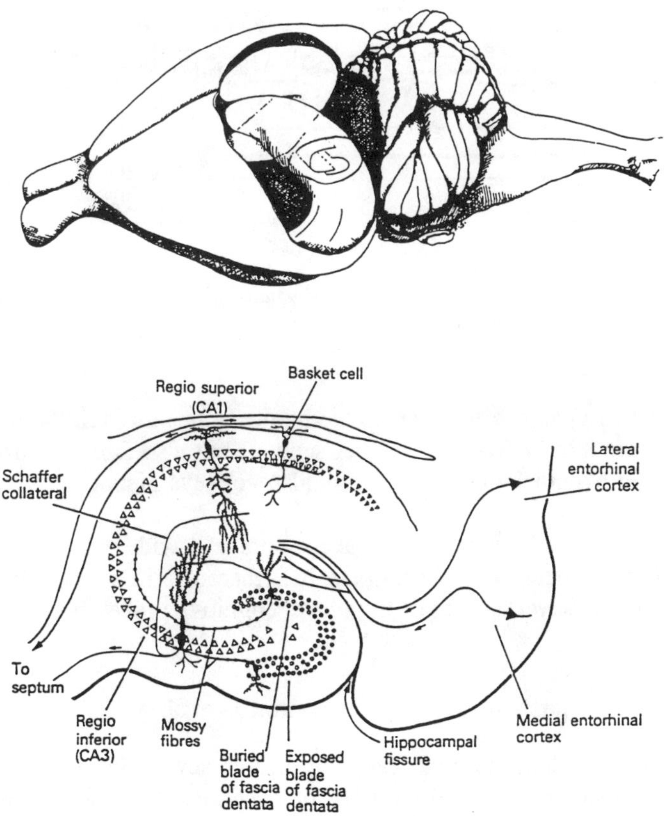

Fig. 6.1. A: Drawing of the left rat hippocampus with the superficial part of the hemisphere removed; B: Section of the middle region of the hippocampus showing intra-hippocampal connections. (Adapted from O'Keefe and Nadel, 1978; and Miller, 1991.)

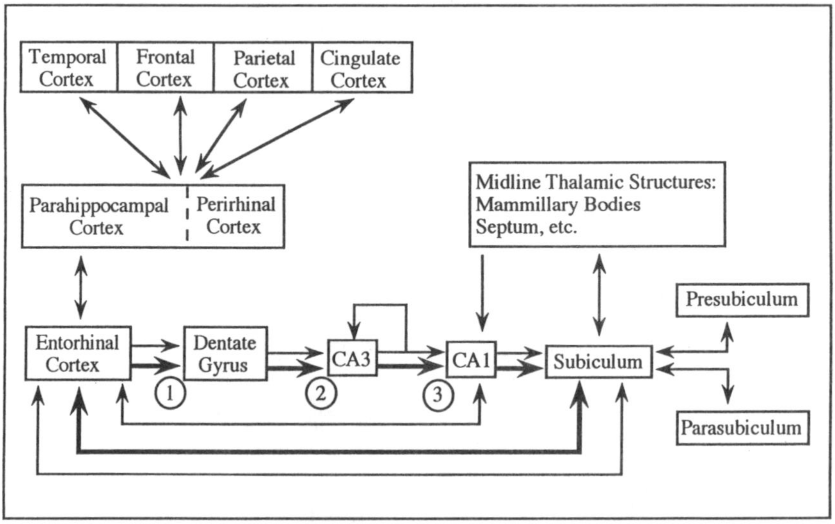

Fig. 6.2. The major intrinsic and extrinsic connections of the hippocampus. The trisynaptic loop is indicated by thick lines (adapted from O'Mara, 1995).

Most subcortical afferents access the hippocampus through the fornix whose fibers originate mostly from the medial septum, but also from brainstem nuclei, thalamus, and hypothalamus. The fornix also conveys hippocampal efference to subcortical nuclei.

In summary, the hippocampus is at the core of a wide network of reciprocal subcortical and neocortical connections. It is therefore in a strategic position to play a key role in several types of high level cognitive processing.

2. The Hippocampus and Traditional Spatial Learning Tasks

Lesions of the hippocampal formation induce severe and permanent deficits in spatial orientation in a wide variety of situations (Rasmussen, Barnes, and McNaughton, 1989). I shall not review here the several hundreds of studies dealing with this issue (for reviews, see Barnes, 1988; Foreman and Stevens, 1987; Nadel, 1991; O'Keefe and Nadel, 1978). Let us just remind ourselves that spatial deficits following hippocampal damage have been observed in the spatial version of the Morris navigation task, with the hidden platform, but not in its nonspatial version in which the platform is visible or signalled by a conspicuous beacon (Kelsey and Landry, 1988; Morris, Hagan, and Rawlins, 1986; Morris, Garrud, Rawlins, and O'Keefe, 1982; Morris, Schenk, Tweedie, and Jarrard, 1990; Schenk and Morris, 1985); in the cross maze (O'Keefe and Conway, 1980); and in the radial arm maze (Jarrard, Okaichi, Steward, and Goldschmidt, 1984; Olton, 1982; Walker and Olton, 1979).

With the exception of the studies quoted above where lesioned rats were subjected to both spatial and nonspatial versions of the Morris navigation task, there are only a few "dissociation" experiments which have been systematically conducted in order to contrast the effects of hippocampal damage in various kinds of tasks (cf. Barnes, 1988, for a review). Whereas lesions of the hippocampal formation in the rat disrupt performance in various tasks requiring the localization of an inconspicuous goal on the basis of extra-maze landmarks (cf. above), such lesions do not affect visual guidance toward a goal in close spatial relationship with a cluster of landmarks in the four arm maze (O'Keefe and Conway, 1980). Similarly, rats with fornix transection show a selective impairment in the retention of a location discrimination but not in the retention of an object discrimination (Olton, Becker, and Handelmann, 1980). Other studies have demonstrated that hippocampal inactivation induces deficits in a delayed alternation task, but not in a reference memory task in which rats are trained to return to the same arm of an apparatus (Barea, Hickey, and Smith, 1988).

A different approach which contributes to the understanding of hippocampal function in spatial cognition consists of studying animals from inbred strains. In this respect, DBA/2 (hereafter "DBA") mice have been found to be impaired in various spatial tasks and exhibit a number of neurobiological alterations at the hippocampal level. On these bases, several authors (Douglas, 1975; Paylor, Baskall, and Wehner, 1993) have considered DBA mice as a genetic model of hippocampal dysfunction. Indeed, intact DBA inbred mice have been found to perform as poorly in spatial learning situations as mice from other strains (e.g., C57BL/6) after a lesion of their normally functioning hippocampus (Ammassari-Teule and Caprioli, 1985; Crusio, Schwegler, and Lipp, 1987; Upchurch and Wehner, 1989).

However, non-spatial discriminations and reversals are also sometimes affected by damage to the hippocampal formation (O'Keefe and Nadel, 1978; Foreman and Stevens, 1987). The conclusion which emerges from O'Keefe and Nadel's analysis of these deficits is that the persistence of inappropriate response habits in hippocampals (for instance making consistent right turns when the positive stimulus is randomly alternated between right and left) accounts for their deficits in non-spatial discriminations and reversals, but not a failure to identify the stimuli per se. Recent data obtained using the dishabituation paradigm (see section 3), which show that hippocampal rats do react to a new object introduced into the test arena, support O' Keefe and Nadel's interpretation. Other theoretical accounts of memory processes in general (e.g., Eichenbaum et al., 1994; Rudy and Sutherland, 1995; see also Chapter 9) provide alternative explanations for some of the non-spatial effects of hippocampal lesions.

In a recent review, Poucet and Benhamou (1996) emphasize a feature which emerges from the abundant literature on the effects of septo-hippocampal lesions on spatial performance, namely that retention of well-learned information is often spared. The most dramatic deficits are observed when new spatial information has to be acquired. For instance, rats with a dysfunctioning hippocampus are capable of navigating to a goal location learned pre-operatively but they are unable to learn a new goal location (McNaughton, Barnes, Rao, Baldwin, and Rasmussen, 1986b; Morris, Barnard, and Biegler, 1994). The sparing of spatial performance is all the more pronounced when there is a long delay between training and surgery (Kubie, Dayyani, Muller, Cohen, Major, & Sutherland, 1990; see also Sutherland, Arnold, and Rodriguez, 1987, for similar effects following dentate gyrus lesions). These data strongly suggest that the role of the hippocampus, which is crucial during learning, becomes progressively less important once consolidation has occurred, i.e., during later stages of the spatial memory process. However, puzzling data obtained by Poucet, Herrmann, and Buhot (1991a) question the continuous involvement of the hippocampus in spatial processing even as early as during acquisition. Rats were chronically equipped with either bilateral cannulae in the ventral hippocampus or a single cannula into the medial septum in order to inactivate these structures by lidocaine injections when necessary.[b] They were trained to locate, in a circular platform with 18 holes at the periphery (Fig. 6.3), the unique hole leading to a hidden shelter in order to avoid bright light (this version of the hole-board task had been designed by Barnes, 1979). In order to control for the possible use of olfactory cues, the board was rotated following each trial, and thus the goal could be located solely on the basis of distal cues. In a first experiment, following 16 days of training at the rate of *one trial per day* without any treatment, the location of the goal was changed on days 17 and 23 and the rats were either sham-injected or injected with lidocaine. Unlike lesions or high frequency stimulation, reversible inactivation of either the ventral hippocampus or the septum failed to induce any deficit in learning the new goal locations, with this distributed trial procedure. In contrast, when rats had to learn a new hole location each day, with 3 massed trials per day (successive trials being separated by a one-minute delay), no improvement within each daily session was observed when the

---

[b] Barnes (1988) has advocated this method which has the advantage of circumventing the problem of synaptic rearrangements (sprouting, reafferentation) which occur after brain damage. It also allows for the functions of the hippocampus to be neutralized at any time in the course of a given behavioral process.

Fig. 6.3. Schematic representation of the circular platform. The platform has 18 holes at the periphery with only one hole connected to a small escape box. (Adapted from Poucet et al., 1991a.)

animals were lidocaine injected[c] (Fig. 6.4). Thus, these data show that a transient inactivation of the hippocampus or of the septum induces a short-term memory deficit. Indeed, ventral hippocampal and medial septal inactivations had already been found to produce working memory deficits in the radial maze (Barea et al., 1988; Mizumori, McNaughton, and Barnes, 1989). In contrast, more striking is the fact that the same inactivation does not interfere with the long-term integration of spatial information, despite the fact that this was presented while the hippocampus or the septum was not functioning. To account for these results, it must be remembered that the lidocaine-injection inactivation procedure disrupted hippocampal functioning only for a short duration (at most 30 min). The authors conclude that hippocampal activity out of the context of learning trials (i.e., during the 23.5 h spent in the home cage) must provide "off-line" processing of the information that was gathered during learning trials, and, in addition, that this would not be required during subsequent trials when the hippocampus is inactivated again.

---

[c] Animals were sham-injected (with saline) or lidocaine injected on alternate sessions.

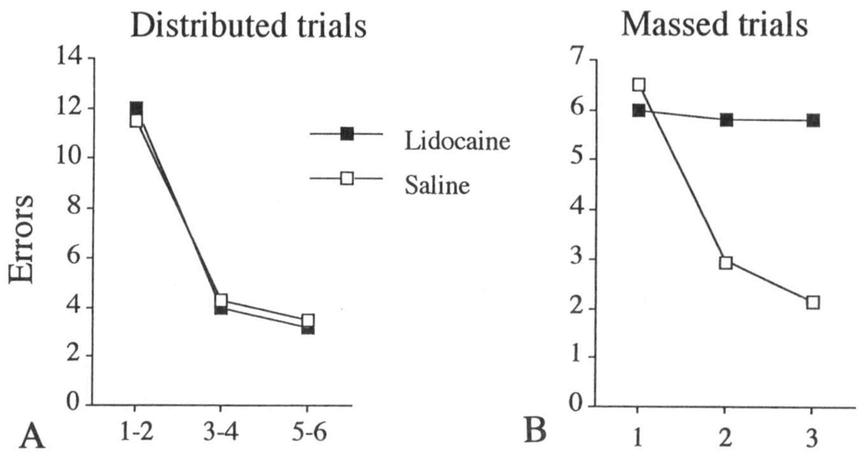

Fig. 6.4. Performance of saline- and lidocaine-injected rats when the trials are distributed (A) or massed (B). (Adapted from Poucet et al., 1991a.)

Interestingly, McClelland, McNaughton, and O'Reilly (1995) have proposed a model of complementary learning systems in the brain which predicts Poucet et al.'s (1991a) data. Memory traces stored immediately after experience in the neocortical areas which have ensured their processing would be reactived in the rat hippocampus during "off-line" periods (corresponding to rest or sleep; Pavlides and Winson, 1989; Wilson and McNaughton, 1994) for consolidation to occur (see Chapter 9).

In summary, a host of data demonstrate that hippocampal lesions in the rat selectively affect spatial performance which relies on spatial representations, whereas they leave guidance behaviors intact. The fact that complex non-spatial tasks may also be affected by such lesions supports the idea that the rodent hippocampus is not only involved in spatial processing, but also in more general memory mechanisms, as in human and non-human primates (for a review, see Eichenbaum et al., 1994).

3. The Hippocampal Formation, Exploration, and Problem-Solving Tasks

The hippocampus (and related structures) is highly involved in exploration and habituation (for reviews, see Foreman and Stevens, 1987; Poucet and Thinus-Blanc, 1992; Thinus-Blanc et al., 1991a). Hippocampal rats are often hyperactive in the open-field but this is not the expression of exploratory activity per se since inter- and intra-session habituation is usually[d] delayed or absent. Similar effects are observed following lesions to the septum (Herrmann, Poucet, and Ellen, 1985a; Poucet, 1989) or medial entorhinal cortex (Schenk, Inglin, and Gyger, 1983).

The behavior of rats with dorsal hippocampal lesions was studied by Save, Poucet, Foreman, and Buhot (1992b) using the reaction-to-change test (see Chapter 2 for details of the procedure). Five objects were scattered in the field. After four sessions of habituation, two out of the five objects were displaced for a test session. In an additional test session, one of the objects was replaced by a new one, occupying the same location. Hippocampal rats failed to react to the spatial change but they were able to explore the new object (Figs. 6.5 and 6.6).

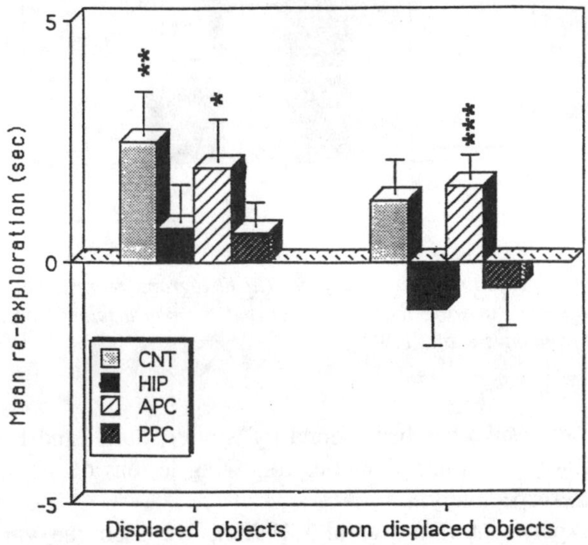

*Fig. 6.5. Behavior in response to the spatial change. The histograms represent the differences in mean (+ S.E.) time spent exploring either the displaced objects or the non displaced objects between post-change sessions and the two preceding sessions for each group. The star(s) indicate(s) a statistically significant difference between pre- and post-change sessions (*** p<.01; ** p<.025; * p<.05). CNT: control group; HIP: hippocampal group. APC: group with a lesion of the anterior part of the associative parietal cortex; PPC: group with a lesion of the posterior part of the associative parietal cortex. (Reproduced from Save et al., 1992b.)*

---

[d] Some exceptions are occasionally observed. For instance, in the experiment by Save, Poucet, Foreman, and Buhot (1992b) presented here, hippocampal rats displayed habituation of both locomotor and exploratory activity.

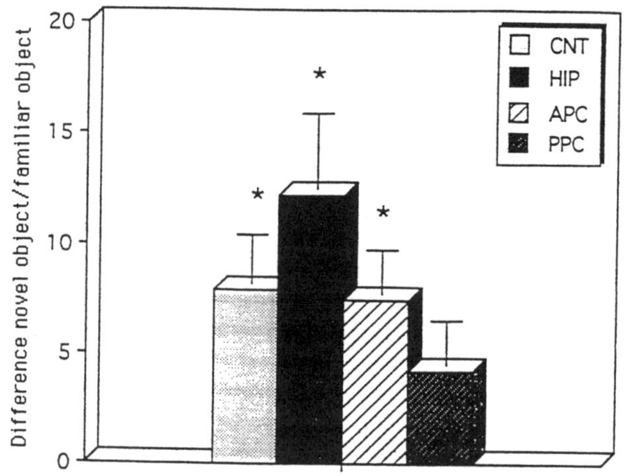

Fig. 6.6. Behavior in response to the object change. The histograms represent the differences in mean (± S.E.) time spent exploring the new object and the non displaced objects, for each group. (Reproduced from Save et al., 1992b.)

A similar pattern of results has been found by Xavier, Stein, and Bueno (1990) and by Poucet (1989) in similar situations following lesions of the hippocampus and of the septum, respectively.

In another experiment (Save et al., 1992a), we used the variant of the reaction-to-change test with a live "stimulus-rat" located under the glass floor of the field during the habituation phase and removed it for the test (see Chapter 2 for details). Rats with bilateral lesions of the dorsal hippocampus showed no habituation of exploratory activity and failed to react to the change, i.e., they did not spend more time above the former location of the stimulus-rat, as did controls.

In that situation, there was one unique visual cue (consisting of a striped pattern, fixed to the inside wall of the field) that could be used to locate the stimulus rat's initial location (since the apparatus was surrounded by white curtains). However, subtle behavioral differences might have been masked by the difficulty associated with the use of such a bare environment. Consequently, in other two experiments, we attempted to make localization of the change easier to detect by adding some cues in the field. Two visually prominent objects were placed beneath the glass floor of the apparatus during habituation and test sessions. These two objects could be either spatially separate from the stimulus-rat location (Dissociated object condition) or close to it (Associated object condition; see Fig. 6.7). In spite of the additional information likely to facilitate the localization of the missing stimulus, hippocampal rats still did not react to the modification even in the Associated object condition (Fig. 6.8).

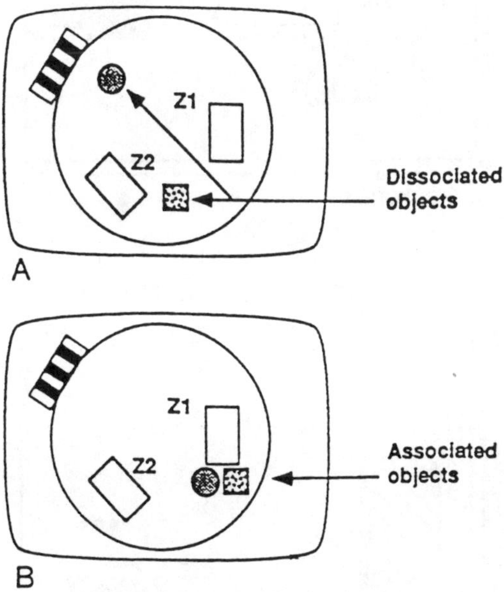

Fig. 6.7. Schematic representation of the apparatus during exploration with objects located under the floor. A: Dissociated object condition; B: Associated object condition. The stimulus rat, when present, was located at Z1. (Reproduced from Save et al., 1992a.)

In a previous experiment conducted in the same apparatus without cue-objects, we (Thinus-Blanc et al., 1991b) had found that the reversible inactivation of the hippocampus by in situ injections of a local anaesthetic, after habituation but before the test session, also resulted in a failure to detect the change. This result allows us to discard the hypothesis that the hippocampal deficit in reaction to a spatial change might result from a failure to encode spatial relations during the pre-test exploratory phase.

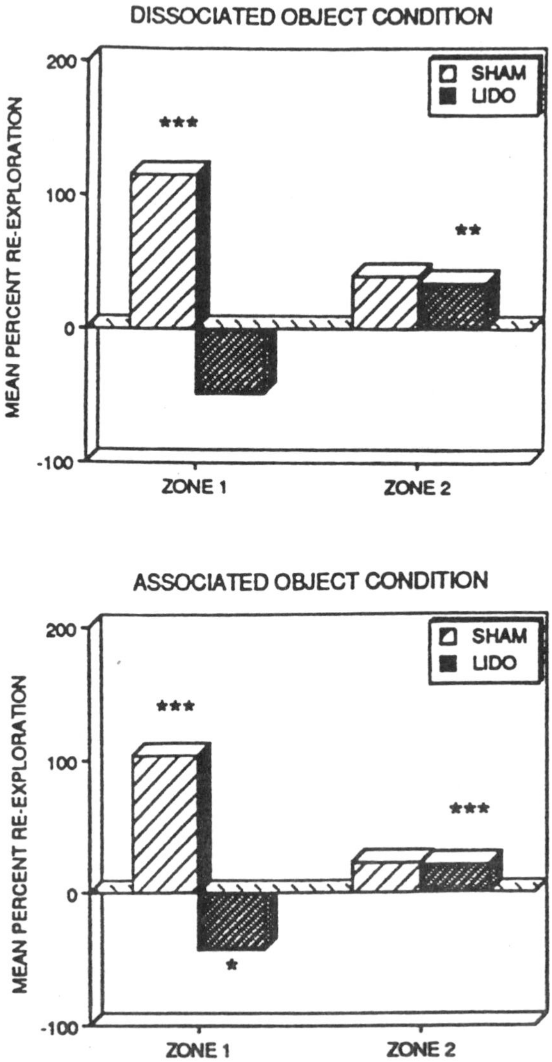

*Fig. 6.8. Behavior in response to the change in the Dissociated object condition (A) and Associated object condition (B). The histograms represent the difference in percentage of time spent investigating the two zones Z1 and Z2 during the test-session (with the stimulus removed but the objects still in the field) and the reference (100%) level of the pretest session (with the stimulus under the floor of the apparatus) for the two groups CNT (controls) and LIDO (lidocaine-injected). The stars indicate a statistically significant difference between the two sessions (*** $p<.01$; ** $p<.025$). (Reproduced from Save et al., 1992a.)*

Variants of Maier's three-table task (see Chapter 2 for details), which relies on exploratory activity, have been extensively used to study the effects of septal lesions on problem solving. For instance, Herrmann et al. (1985a) added runways to the basic Y-shaped apparatus so as to form a triangle which included the Y-shaped maze. During the daily habituation phase, both control and septal groups displayed a preference for the outer runway configuration but, again, septal animals' exploratory activity did not significantly decrease over time. During testing, however, septal rats retained their preference for outside paths and performed around chance whereas normal subjects attempted a solution by using the inner runway configuration.

In another experiment (Poucet et al., 1990), septal and control rats were tested in a variety of runway configurations which linked the three tables (cf. Fig. 2.10). In all variations of the task, septal rats were impaired on test trial performance and displayed stereotypical body turn responses.

Finally, with the three tables interconnected by the classical Y-shaped runway system, Poucet and Herrmann (1990) used two experimental situations aimed at decreasing the difficulty of the spatial reasoning task. In a fixed-goal (FG) situation, the food was located on the same table throughout testing, and the start table was randomly varied from day to day. In a variable-goal (VG) situation, the start table remained constant but the food was randomly distributed on one or the other of the two remaining tables. In both tasks, normal animals performed better than hippocampal rats whose performance, however, improved over days in the FG but not in the VG task (Fig. 6.9). As previously, septal rats did not display habituation of exploratory activity.

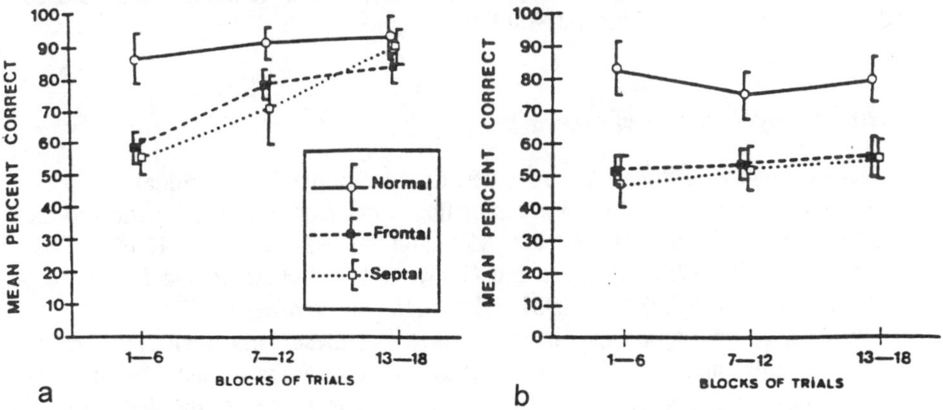

Fig. 6.9. Mean (± S.E.M.) Trial 1 performance (% correct) in the Fixed-goal (a) and Variable-goal (b) tasks over 6-day blocks (reproduced from Poucet and Herrmann, 1990).

Taken together these data have led to the conclusion that septal damage interferes with the building of comprehensive spatial representations. As a consequence of this impairment, septal rats seem to rely upon "taxon"-type strategies that can combine guidances and orientational responses (see Chapter 1, section 4).

In a recent study, DBA mice were compared with C57BL/6 mice in a reaction-to-change test (Ammassari-Teule et al., 1995). We found that C57 mice displayed strong exploratory reactions to the displaced objects whereas DBA mice did not consistently react to the new arrangement. Nevertheless, both strains reacted to the substitution of a new object for a familiar one at the same location. Indeed, intact DBA mice's pattern of results mimics the effects of hippocampal lesions in rats (see above) and C57 mice tested in similar situations (Thinus-Blanc, Save, Rossi-Arnaud, Tozzi, and Ammassari-Teule, 1996c).

In summary, the hippocampal formation appears to play a key-role in the organization of new incoming information. During exploratory behavior different types of investigative movements (eye and head scanning movements, rearing up, and so on) generate ever changing sensory stimulation. In this respect, one hippocampal function during exploration could be to ensure the short-term storage of sequentially acquired pieces of information about the environment (e.g., various local views of the same place viewed from different perspectives) matched with the subject's displacements. Indeed, this short-term storage is necessary in order, for instance, that various slightly different and contiguous local views are placed into the same category (i.e., belonging to the same particular place). The lack of habituation observed in hippocampal animals may be due to the fact that new spatial links between sequentially perceived environmental features cannot be stored in memory, even if non-spatial features are processed and memorized (Thinus-Blanc, 1988b).

4. Hippocampal Place Cells

Single unit recordings have revealed that a number of pyramidal complex-spike cells in the CA1 and CA3 areas of the hippocampus fire in relation to the rat's location within the environment (McNaughton, Barnes, and O'Keefe, 1983; Muller and Kubie, 1987; O'Keefe and Conway, 1978; O'Keefe and Dostrovski, 1971; O'Keefe and Speakman, 1987; Olton, Branch, and Best, 1978).

In order to investigate the firing properties of these cells, many authors, in particular Muller and coworkers, have used a circular arena placed in a cue-controlled environment. A large conspicuous cue-card is affixed to the wall. Before recordings are made, hungry rats are trained to retrieve small food pellets scattered into the chamber. As a result, the rats spend almost all of their time walking about, and they visit all parts of the arena. A camera above the cylinder records rats' trajectories. In addition to the recording headstage, two light-emitting diodes (LEDs) of different colors are arranged on the rat's head so as to permit calculation of the animal's head (Fig. 6.10). Thus, by sampling and computing the coordinates of each LED, it is possible to match the electrophysiological activity with both the location and direction of the animal's head (for further details, see Muller, Kubie, Bostock, Taube, and Quirk, 1991b).

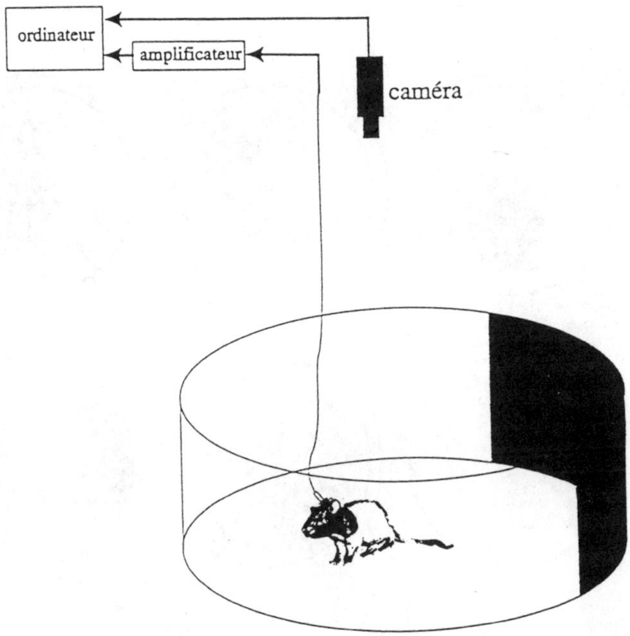

Fig. 6.10. Diagram of the device for recording place-cell activity from the hippocampus in freely-moving rats.

Place-cell firing fields are stable over time (Best and Thompson, 1989; Muller, Kubie, and Ranck, 1987), persist if some of the environmental cues are removed (O'Keefe and Conway, 1978; O'Keefe and Speakman, 1987) and have directional correlates if the rat's displacements are constrained along determined paths (as in the radial maze) (McNaughton et al., 1983).

Until recently all place-cell recordings had been made from the dorsal hippocampus, a region where CA1 and CA3 fields are easier to access for practical reasons. However, Poucet, Thinus-Blanc, and Muller (1993, 1994) and Jung, Wiener, and McNaughton (1994) have discovered place-cells in the ventral hippocampus, the properties of which appear to be rather similar to those of place-cells recorded from the dorsal hippocampus. In all cases, the "firing fields" of these cells (i.e., the region of the apparatus in which they fire) is independent of the particular orientation of the subject but relies on the environmental configuration of cues. If the latter is rotated, the firing field is also rotated to the same extent (Fig. 6.11).

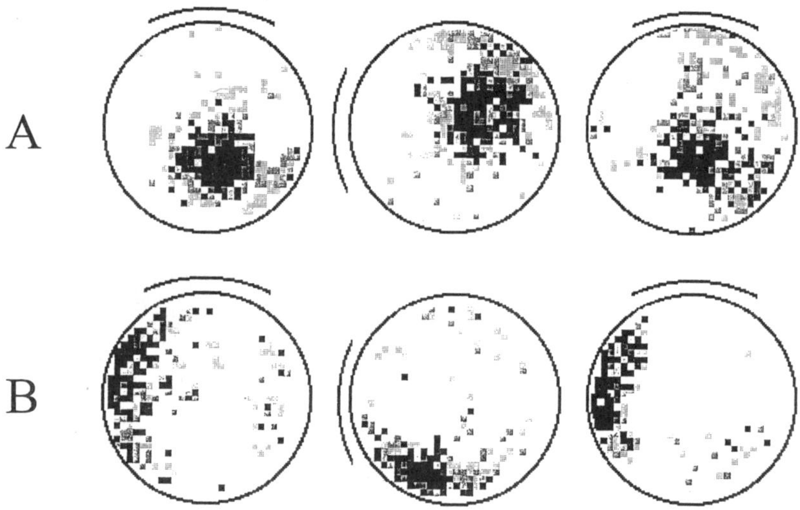

Fig. 6.11. The effects of cue rotation on the spatial firing pattern of a ventral (A) and a dorsal (B) place cell. The 90° clockwise rotation of the cue-card (indicated by an arc outside the cylinder) produced equal rotation of the place field.

One of the main questions which has been addressed so far concerns the nature of the information that controls place-cell activity. A number of studies have demonstrated the importance of visual cues either inside the recording arena, such as the cue-card represented in Fig. 6.10, or visual cues that are outside the apparatus but which are relatively proximal: A set of cues distributed within a cue-controlled box (O'Keefe and Speakman, 1987) had a strong effect on place-cell activity since the rotation of the cue(s) induced a corresponding rotation of the firing-field (for a review, see Muller et al., 1991b).

Several attempts (e.g., Poucet, Cressant, and Muller, 1995) have been made to examine whether objects placed in the field have the same properties as the cue-card affixed to the limiting wall, or extra-apparatus landmarks, in terms of their ability to "polarize" the situation and drive place-cell activity. Even when objects were placed in the field as early as the first minutes of the animal's initial familiarization with the apparatus, firing fields, when they were observable, did not "follow" the object displacements. However, Cressant, Poucet, and Muller (1996) have recently discovered that if, instead of scattering the objects in the central region of the experimental arena, they are simply placed in close proximity to the surrounding wall, or they are pooled so as to form one large, unique compound object, the objects controlled the activity of place-cells and their corresponding firing fields as accurately as a peripheral cue-card. The rotation of the object configuration or the displacement of the compound object induced an equivalent displacement of the firing-field.

O'Keefe and Burgess (1996) have identified other environmental features that control the location and shape of place-cell firing-fields by recording from the same cell in 4 rectangular boxes which differed solely in the length of one or both sides (a small square, a large square, a horizontal rectangle and a vertical rectangle). They found that the walls bounding the environment controlled the location and shape of the firing fields in various ways. Some fields were observed at a fixed distance from a wall, others peaked at a fixed proportion of the distance between two walls. The authors also found fields the shape of which varied systematically between boxes differing in length along one direction; the fields were stretched or pulled apart along that direction in the larger box.

Thus, several types of external information seem to control place-cell activity as reflected by firing fields' "reactions" to environmental manipulations: visual intra- and extra-apparatus cues that do not modify locomotor activity, objects within the field (provided that they are in close proximity to the wall or that they form a large, unique object), and the size and shape of the bounding walls, that can be visually and non-visually perceived and which, in addition, direct trajectory kinematics. Although place-cell activity depends on the configuration of visual (and non-visual) cues, these cells, as emphasized by Muller et al. (1991b), are not sensory but represent an encoding of the animal's absolute location within the environment (O'Keefe and Nadel, 1978).

Among these various types of information, those provided by vision appear to play a prominent role in driving place-cell activity. In her doctoral dissertation, M. Krimm (1994), adopting a developmental perspective, has examined the effects of early visual deprivation on place-cell activity. Her question was: is early vision necessary for place-cells to acquire spatial encoding properties, and are these properties similar as those found in light-reared animals? Rats were kept in total darkness during the first two months of life. They were then placed in normal lighting conditions for 1.5 month, and then equipped with the head-stage for recording hippocampal place-cell activity. A cylindrical chamber as that shown in Fig. 6.10, with a cue-card taped to the boundary wall, and food pellets scattered on the floor, was used. She found place-cells whose firing had nearly the same characteristics (e.g., frequency of discharges) as those of light-reared animals. However, when the cue-card was rotated, only 4 out of the 17 cells which were recorded had a firing field which followed the cue rotation. The firing fields of the 13 remaining cells were considerably altered by the cue-card manipulation, but were restored when the cue-card was replaced at its initial location. Since early visual deprivation, followed by visual experience when adult, did not eliminate place-cell activity but only modified responses to environmental manipulations, we examined the possibility that the properties of these cells had become established later on, when animals had been exposed to normal illumination. For that purpose, hippocampal activity has been investigated in adult rats that were enucleated a few days after birth. The objects used previously by Cressant et al. (1996) were placed in the cylinder. Preliminary results indicate the existence of place-cells, the firing-fields of which have roughly the same properties as those observed in light-reared animals. In addition, the same alteration of the firing-field is observed when the cue-object is rotated, in nearly the same way as with dark-reared animals (Save, Cressant, Thinus-Blanc, and Poucet, 1996a). These results, if they are confirmed by additional recordings, demonstrate that although it is important in the control of place-cell activity, vision, and early visual experience in particular, is not necessary for the setting up of place-cell activity.[e] Nevertheless, it remains for us to account for the alteration of the firing-fields when intra-apparatus cues are manipulated. One possible explanation is that modifying proximal cue location, leaving the remaining environmental framework otherwise stable, amounts to a conflict situation. Whereas light-reared animals' place-cells up-date the modification by reference to the stable environment, those of visually-deprived rats may lack this plasticity and react to the conflict by incoherent firing. Further experiments and recordings are required in order to test this hypothesis.

---

[e] Early or congenital blindness induces severe spatial impairments in humans (for a review, see Thinus-Blanc and Gaunet, 1996), and to a lesser extent in animals when the latter can recover sight when adults (Krimm, 1994; Thinus-Blanc and Krimm, 1993). One possibility is that early visual experience would have a calibrating role for non-visual information during development, in particular for spatial processing. This hypothesis would account for the spared spatial abilities of late-blinded people. In line with this hypothesis, place-cells which play an important role in spatial encoding, could be affected by early visual deprivation.

Besides the role of information provided by the environment (external information), another type of input has been found to affect place-cell activity, namely movement-generated information while the experimental animal is moving around in space (see Chapter 5). For instance, when recording from the CA1 field of the hippocampus, Wiener, Paul, and Eichenbaum (1989) have found that the firing rate of place-cells varied systematically in relation to the speed, direction and turning angle while the rat was moving through the place-field. However, these data could be interpreted as reflecting a strong modulation of place-cell firing by locomotor activity regardless of spatial encoding. A more convincing interpretation in spatial terms has been provided by Quirk, Muller, and Kubie (1990), who showed that the firing of place cells is not disrupted by turning off the light when the rat is already been placed in the recording chamber. By contrast, when the rat is placed in the recording chamber in darkness, there is a strong modification of the patterns of firing fields. The authors propose that after the light has been turned off, rats may keep track of their position with vestibular and proprioceptive information. However, it is unlikely that this information could be sufficiently accurate for maintaining a constant 'fix' on the animal's successive positions within the experimental arena throughout a long recording session because of the cumulative error of this system in the absence of external feedback.

Recently, data have accumulated that demonstrate the influence of movement-related information on place-cell activity. Sharp, Blair, Etkin, and Tzanetos (1995) have provided evidence that not only vestibular activation but also visual motion (movement-induced optic flow) control place-cell firing fields. Place cells were recorded as animals locomoted in a circular apparatus that was made so that its walls (painted with vertical white and black stripes) and floor could be independently rotated, to provide visual motion and vestibular inputs, respectively. Different patterns of modification of the firing fields were observed according to the various combinations of visual motion and vestibular activation. For instance, fast rotations of the floor (vestibular activation) caused fields to fire in a rotated position in relation to the apparatus floor; this effect, which reflects an up-dating of the animal's position, was all the more likely to occur when vestibular and visual motion manipulations were made simultaneously and were matched (with the same direction and angle of rotation). In contrast, when the two types of manipulation were conflicting (different directions and angles of rotation), the overall firing characteristics were altered, or cells stopped firing altogether. Other recent studies in the same vein have provided additional support for the idea that the hippocampus accurately detects passive movements and can also update the animal's position by reference to stable features of the environment (Knierim, Kudrimoti, and McNaughton, 1995; Wiener, Korshunov, Garcia, and Berthoz, 1995).

In view of these data, it is not surprising that place-cells have been found to be active during exploration (Muller and Kubie, 1987; Muller et al., 1987; O'Keefe and Nadel, 1978) which consists of many back-and-forth movements between objects and in the collection of environmental information from multiple sources. O'Keefe (1976) had already made a distinction between place units (see above) and "misplace" units which fired maximally, at a particular place, when, for instance, stimuli had been removed from the environment. However, the question concerning the time necessary for firing fields to become established is still a matter of debate. Ranck (1973) reported the existence of multimodal neurons which habituate rapidly to repeated stimulus presentation but are responsive to novelty, with some cells firing during orienting behavior. More recently, Muller and Kubie (1989) have shown that when rats are exposed to a new stimulus situation in an open-field, place-cells become progressively active while the rat is at a given location in relation to a visual cue. Conversely, according to Hill (1978), no learning period appears to be required for spatially selective firing to occur; the spatial selectivity of place-cells appears not to depend on prior experience of the rat within the environment (McNaughton, Barnes, Meltzer, and Sutherland, 1989a). Lastly, Wilson and McNaughton (1993) found rapid changes to occur in the firing properties of neurons that initially fired in the familiar half of a rectangular recording chamber, when animals had access to and explored the other half which was novel for them.

Despite discrepancies between data showing immediate or progressive firing of place units in an unfamiliar situation, the important point is that place-cells are found to be active during exploration. Another finding which emphasizes the active feature of exploration and the involvement of the hippocampal formation is that spatially selective firing of hippocampal cells is totally removed by head and/or body restraint (Foster, Castro, and McNaughton, 1989). Finally, some characteristics of dentate granule cell activity (such as the growth in synaptic strength, excitatory post synaptic potential and increase in the population spike components of the response) are specifically related to the exploration of a new environment (Green, McNaughton, and Barnes, 1990).

In summary, although place-cells are not strictly sensory-bound (they continue to fire in a lack of visual stimulation), they are controlled by environmental features such as particular types of landmark, geometric features, and so on. Accordingly, their activity reflects the integration of movement-generated information which allows for the up-dating of the animal's position by reference to a stable environment.

5. "Head Direction" Cells

Other kinds of cell, "directional" or "head-orientation" cells, have been discovered by Ranck (1984, 1985), and Taube, Muller, and Ranck (1990 a and b) in the post-subiculum, posterior cortex (Chen, Lin, Barnes, and McNaughton, 1994a; Chen, Lin, Green, Barnes, and McNaughton, 1994b), thalamus (Taube, 1992; Mizumori and Williams, 1993; Knierim et al., 1995), and striatum (Wiener, 1993). Unlike place-cells, the firing of these cells is not related to the rat's current location but only to the direction in which the head is pointing. Therefore, a cell whose preferred direction is, say, the east, fires in any place of the cylinder provided that the head points toward the east; the vectors in the direction of the maximal firing are parallel throughout the apparatus. Taube et al. (1990 a and b) have shown that post-subiculum head-direction cells fire only when the rat's head points in an approximately 90° range of angles in the horizontal plane. For each individual cell, preferred directions occur with equal frequency in each of the four quadrants of the circular apparatus. Different cells have different preferred direction in all azimuthal directions. Although they are not "sensory", the preferred direction of these cells depends on the cue inside the recording arena. The rotation of the cue card produces near-equal rotation of the preferred firing direction of the head-direction cells. However, the cue card is not the only modulating factor that influences firing rate since its removal does not disrupt the firing fields for preferred directions (Mizumori and Williams, 1993).

As with place-cells, head-direction cells can use movement-generated information, particularly in the absence of visual landmarks. However, when there is a conflict between internal and external information, sensory inputs on to head-direction cells from external landmarks override idiothetic input (Goodridge and Taube, 1995; McNaughton, Markus, Wilson, and Knierim, 1993b). Interestingly, these data are consistent with Etienne et al.'s (1990b) findings; these authors showed that when there is a conflict between information from internal and external sources, the latter predominates over the former.

In addition to HD and place cells, the hippocampal formation contains neurons whose spatial signals are more complex. For example, cells from medial entorhinal cortex (Quirk, Muller, Kubie, and Ranck, 1992), dentate gyrus (Jung and McNaughton, 1993) and subiculum (Muller et al., 1991b) act like place cells although their firing fields are usually less specific than those of hippocampal place cells. Lastly, cells found in the subiculum have both locational and directional correlates (Muller et al., 1991b).

6. Concluding Remarks

A remarkable feature which emerges from the overview presented above is the convergence of findings emanating from various levels of approach.

Following lesions in the septo-hippocampal system, impairments are observed at all stages of spatial information processing, at least within a limited time window: when an animal is first exposed to an unfamiliar situation (lack of habituation), when a comparison process is required (failure to react to a spatial change), and in traditional spatial learning tasks (e.g., large number of errors in the Olton maze, longer latencies in the Morris navigation task). Deficits are more pronounced when short-term memory is involved, as in the radial maze, the Maier three-table test, or in the hole-board situation (Poucet et al., 1991a), in the latter case, when the location of the goal changes each day. Hippocampal animals are also impaired in spatial navigation tasks, when long term memory predominates; though the goal location does not change during learning, it has to be deduced from information provided by the environmental array. However, in such cases, it has been observed that preoperatively trained animals are still able to display accurate navigation (Morris et al., 1994), especially if the delay between learning and surgery has been long (Kubie et al., 1990). These data lead to the suggestion that the hippocampal formation is not the site of long-term storage of spatial memories per se. The issue of the permanent versus temporally graded role of the hippocampus in spatial memory (and in memory in general) is highly controversial. Even more controversial is the conclusion that can be drawn from Poucet et al.'s (1991a) study, namely that the hippocampus could process "off-line" spatial information during the acquisition phase. Obviously, this conclusion implies that while the hippocampus is inactivated, some other structures could, perhaps temporarily, hold information which would later transfer to the hippocampus to be processed.

More generally, such data stimulate the debate about whether the hippocampus is involved at all stages of spatial processing and memory, or whether it plays a transient role which could be assisted in early stages of spatial processing (as early as the acquisition phase), and perhaps totally overridden at later stages by other brain areas (see Chapter 8).

The types of cell that have been found in the hippocampus and associated structures, as well as the firing characteristics of these units, bring other arguments in support of a spatial function for the hippocampal formation. Indeed, the latter contains all the basic ingredients for spatial processing to be performed. At the psychological level, the importance of external and movement-related information in spatial knowledge has been emphasized in preceding chapters (see Chapters 2 and 5): place-cells' and head-direction cells' firing has been found to depend on both types of information. Some of the psychological models presented in Chapter 5 involve local processing of landmarks and the need for an overall reference or compass, for example Collett et al.'s and Poucet's models (see Chapter 5): place-units appear to be good candidates for the former type of processing whereas head-direction cell properties seems to be the neurobiological substrate of the second. Other types of unit such as those found in the subiculum, which have both locational and directional correlates (Muller et al., 1991b), might reflect the integration of both types of information. This diversity of spatial encoding among hippocampal cells, together with the effects of lesions, has prompted the elaboration of various models which are discussed in Chapter 9. Some of these theoretical accounts might be termed "hippocampus-centered" inasmuch as that they assume specific spatial processing to be exclusively performed by the hippocampal formation. In contrast, other models reflect a more comprehensive view of the neural substrates of spatial processing and advance, more or less explicitly, the notion of distributed spatial processing and memory across neocortical areas, in connection with the hippocampal formation. Indeed, some of the data presented in the following two chapters which deal with the spatial function of the parietal, prefrontal and occipital areas bring experimental support to the idea that spatial processing is not exclusively confined to the hippocampus.

CHAPTER 7

THE ASSOCIATIVE PARIETAL CORTEX: AN INTERFACE BETWEEN THE PHYSICAL WORLD AND REPRESENTATIONS?

"Maybe, there is no neural representation of physical space as such. Rather, the 'real' space that common sense persuades us to believe in operationally may be an emergent property of networks of neurons, which are particularly concentrated in the parietal cortex." (Stein, 1991, p. 210)

It is unquestionable that the hippocampus plays a key role in spatial processing. Examples of actual spatial performance that are spared by hippocampal lesions are rare. The deficits are usually pronounced and unrecoverable. Moreover, data from electrophysiological studies add further support for a spatial function for the hippocampus. Does that mean that the hippocampal formation is the unique site in the brain where spatial processing is achieved? The question can be formulated in another way: are there aspects of spatial processing that are *not* ensured by the hippocampus but that are necessary in order that the whole process can take place, from the very first beginnings of exploration up to the point where orientation is based on coherent, effective representations? It appears difficult to maintain that a single structure, what is more, a paleocortical structure, acts alone at all the various stages of the complex spatial process. It is common sense to assume otherwise.

Other less intuitive arguments offer additional support to the notion of cooperation and interaction among brain regions. First, several hypotheses have been proposed recently to account for memory in general via a cortico-hippocampal interplay (see Chapter 9). Second, in humans, spatial disorders are usually associated with posterior parietal cortex damage. As a matter of fact, though it is far less developed than in primates, rats also have a parietal cortical area. What is its function? Is it radically different from at least some aspects of what are supposed to be posterior parietal functions in humans? Similar considerations apply to areas of prefrontal cortex.

In the present chapter, the spatial functions of the associative parietal cortex are examined. Several studies presented in this chapter (and in Chapters 6 and 8) aimed to dissociate the effects of lesions of various neocortical regions and those of hippocampus. However, this approach has not been conducted systematically.

Therefore, for the sake of clarity, the spatial function of each of these structures will be considered separately (the possible interactions between neocortical regions and the hippocampus are discussed in Chapters 9 and 10). Some of the effects of parietal lesions on sensorimotor behaviors are described in order to examine the possibility that genuine spatial behavioral deficits may be a consequence of impairment occurring "upstream" in spatial processing.

1. Anatomical Recall

The neuroanatomical boundaries of the associative parietal cortex (APC) remain unclear and the areas defined as APC vary between authors. Kolb and Walkey (1987), using retrograde labelling, delineated an area which could be structurally analogous to the posterior parietal cortex of primates on the basis of its connectivity with other structures. Lesions in this area resulted in retrograde degeneration in the dorsolateral and lateroposterior nuclei of the thalamus. It was concluded that this region is likely to be homologous with Krieg's area 7 (1946) in primates. Nevertheless, different authors have defined the APC as a more anterior region to that described by Kolb and Walkey (DiMattia and Kesner, 1988 a and b; Kesner, Farnsworth, and DiMattia, 1989).

Degenerative patterns have been found in the same thalamic nuclei however the lesion was made. That is probably due to the fact that these two localizations overlap. Unfortunately, there is a shortage of detailed studies of degeneration in thalamic nuclei following discrete parietal lesions; but the possibility that there are various subregions of the APC which subserve distinct functions should not be discarded (see Save, 1993, for further discussion).

The APC of the rat has connections with primary and secondary visual areas Oc1 and Oc2 (Miller and Vogt, 1984; Torrealba, Olavarria, and Carrasco, 1984; Kolb and Walkey, 1987) and with somatosensory and frontal cortices (Kolb and Walkey, 1987). Fig. 7.1 represents the main cortico-cortical connections of the APC. Among subcortical relationships, recent neuroanatomical studies have shown that there are monosynaptic connections between the vestibular nuclei and the APC in the rat (Guldin, Mirring, and Grüsser, 1992). It should be emphasized that no direct connections with the hippocampus have been demonstrated so far.

2. The Effects of APC Lesions on Motor, Visual, and Visuomotor Behaviors

APC injuries have more or less pronounced effects at various stages of spatial and non-spatial processing and are not restricted to complex navigational spatial tasks.

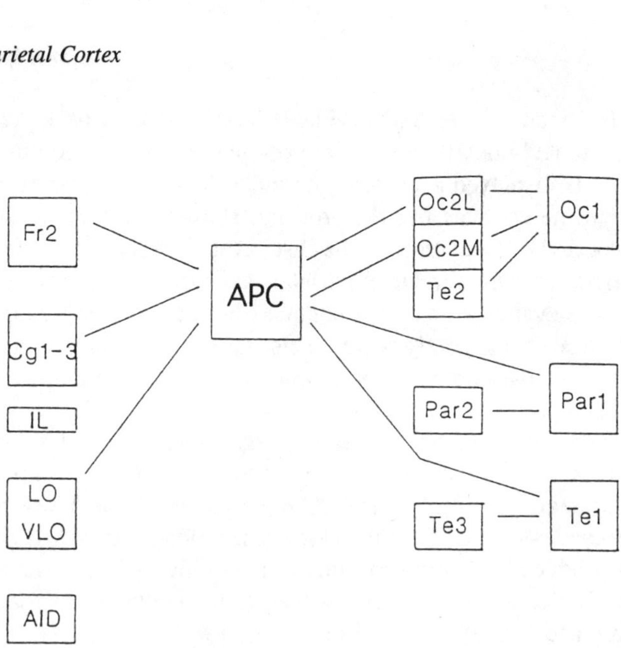

Fig. 7.1. Long cortico-cortical connections of the primary and secondary sensory cortices to the APC in the rat. Abbreviations are those proposed by Zilles (1990). Ocl, Oc2M, Oc2L: occipital cortex; Parl, Par2: somatosensory parietal cortex; Tel, Te2, Te3: temporal cortex; Fr2: frontal cortex; Cgl-3: cingulate cortex; LO, VLO: orbital prefrontal cortex. (Reproduced from Kolb, 1990a.)

The motor impairments which have been observed following APC injuries are mild, transient and limited to the placement of posterior paws while walking on an elevated narrow flat beam (Kolb and Whishaw, 1983; Kolb and Walkey, 1987). At the perceptual level, APC lesions affect neither simple visual discrimination (McDaniel and Thomas, 1978; Kolb, Buhrman, and McDonald, 1989), nor the discrimination of geometrical shapes (Boyd and Thomas, 1977; McDaniel, Wildman, and Spears, 1979), nor visual recognition (Kolb et al., 1989). A lack of any deficit is also observed when rats have to discriminate between patterns in different orientations (McDaniel and Brauch, 1984). In contrast, McDaniel and Wall (1988) have provided evidence that APC lesions induce deficits in the retention of a black/white discrimination, and in the acquisition and retention of obliquely oriented patterns. This impairment may be due to the fact that the damage was located at the posterior part of the APC (and perhaps encroached upon the occipital visual areas) since the same authors failed to demonstrate a deficit with more anterior lesions.

The patterns of locomotion toward visual targets in a runway have been found to be less accurate in rats with bilateral APC damage than controls and unilaterals (Foreman, Save, Thinus-Blanc, and Buhot, 1992). When flashing lights were presented unexpectedly in the peripheral visual field during their run, bilateral parietals displayed a tendency to make fewer orienting responses to the distractors than the controls but they did not show total neglect. After habituation to the repeated presentation of the distractor, effects due to the omission of expected distractors were similar in parietals and controls: stimulus omission per se produced no elevation of running latencies but had in both groups a profound effect on the trial immediately after. This pattern of results suggests that the APC influences the parameters of orientation but is not essential for the initiation of orienting responses.

In spatial tasks requiring egocentric organization, the deficits consequent upon APC injuries are diversified according to the difficulty of the task. For instance, Kesner et al. (1989) trained APC rats in a variant of the radial maze: rats had to visit the alleys adjacent to the starting arm where they had been released. APC lesions produced no deficit on this task. Similarly in the water maze, with the platform visible, parietal rats learned as well as controls to reach the escape platform (Kolb and Walkey, 1987; Kolb et al., 1989).

Davis and McDaniel (1993) trained rats in a visuo-spatial conditional problem. A circular swimming pool was modified by adding partitions which defined a T-maze. At the end of the start arm (i.e., at the choice point), a white or a grey card cued that the escape platform was located at the end of the right or the left alley, respectively. Parietal rats displayed significant deficits in this task. In this experiment, however, animals were free to self correct and thus the error cost was not very great. In a further study, McDaniel and Skeel (1993) used the same procedure but with a greater response cost: animals were confined to the incorrect arm for 15 s prior to being allowed to escape to the platform. In addition, stimulus salience was increased. In that case, no deficit was observed in the parietals.

In a recent experiment, McDaniel, Via, Smith, Wells, Fu, Bishop, Boyd, and Ledesma (1995) trained APC rats for response learning in a cross shaped swimming maze (with distal visual cues available). Two opposite starting points were used and the animals had to make the same turn whatever the starting point from which they were released. They were not confined to the incorrect arm if they made an error and could immediately self correct. The analysis of behavior revealed that animals with either bilateral or unilateral right APC lesions committed significantly more errors than did either the control or the left parietal injured rats.

In summary, APC lesions do not seem to dramatically affect gross visual, motor, or visuomotor abilities. In contrast, when a stimulus-response association has to be established and used, deficits are observed in a simple situation (a T-maze) so long as the cost of errors is low or when the task involves two distinct S-R associations (as in the cross maze).

3. The Effects of APC Lesions on Reactions to Spatial Novelty

The behavior of rats with APC injuries has been studied by Save et al. (1992b) and compared with that of hippocampal animals (see Chapter 6) using the reaction-to-change test with object scattered in the field. Two types of bilateral parietal lesion were made. The first one corresponded to the anterior part of the APC (APC-ant), that was defined by Kolb and Walkey (1987), whereas the other one, more posteriorly placed (APC-post) was the same as that made by Kesner and collaborators. All of the parietal rats displayed a normal pattern of habituation. In contrast, reaction to the spatial change (two out of the five objects were displaced) was observed in the APC-ant group only (Fig. 6.3). However, unlike control animals which selectively re-explored the displaced objects, the APC-ant rats reinvestigated all of the objects as if they could not identify the origin of the spatial modification. The APC-post rats did not react to the spatial change. When a familiar object was replaced by a new one, all of the animals reacted selectively to the modification (Fig. 6.4).

The behavior of APC lesioned rats has also been compared with that of hippocampal rats (see Chapter 6) in the variant of the reaction-to-change test with a live "stimulus-rat" located under the glass floor of the field (Save et al., 1992a). Parietal lesions were located in the anterior part of the APC. Again, rats with lesions displayed normal patterns of habituation, but, unlike the situation with the reaction-to-change test with objects, reported above, they did not react to the change, i.e., they did not spend more time at the location where the stimulus-rat was initially placed.

When two objects were placed under the glass-floor (Dissociated object and Associated object conditions, Fig. 6.5), in order to facilitate the localization of the missing stimulus, parietal rats did not react to the modification (Fig. 6.6). However, a selective preference for staying at the stimulus-rat's previous location was found in parietal animals as was also seen in controls in the Associated object condition.

In summary, in spite of normal patterns of habituation, rats with parietal lesions do not always react to a spatial change, particularly when the change consists of removal of a stimulus. When they do react, as in the experiment using objects, their reactions are unselective.

4. The APC and the Maier Three-Table Task

The three-table task is another paradigm, more direct than "reaction to spatial change" tests, in which the behavioral effects of APC lesions can be examined. We (Thinus-Blanc, Save, Poucet, and Foreman, 1996b) recently conducted an experiment in that situation with two groups of rats with anterior and posterior parietal lesions, respectively (APC-ant and APC-post, as in Save, et al.'s [1992b] experiment). As already shown, habituation occurred in both groups of parietal rats. However, the general level of activity, similar to that of controls during the first six-day blocks of the experiment, increased progressively across days (Fig. 7.2). Performance on the tests was found to follow the same course as the activity change. During the first block, all the APC-injured animals performed at about chance level. Then, their performance level increased progressively and reached that of the controls by the end of the experiment (Fig. 7.3).

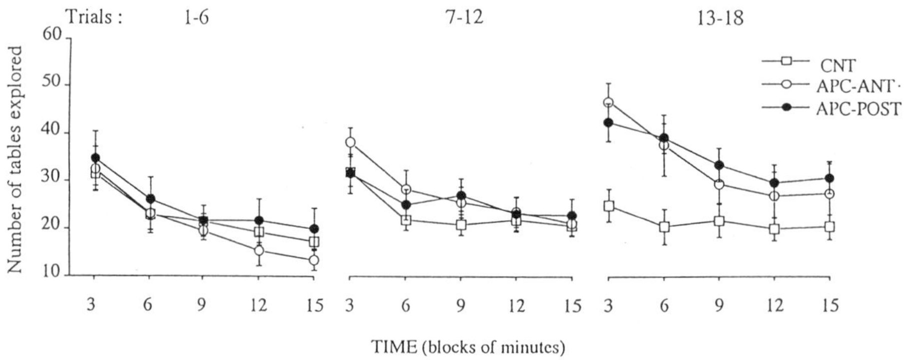

Fig. 7.2. Mean number of tables explored over 3-min intervals in each 6-day blocks of training for CNT, APC-ANT, APC-POST groups. Standard deviations are shown as vertical bars. (Reproduced from Thinus-Blanc et al., 1996b.)

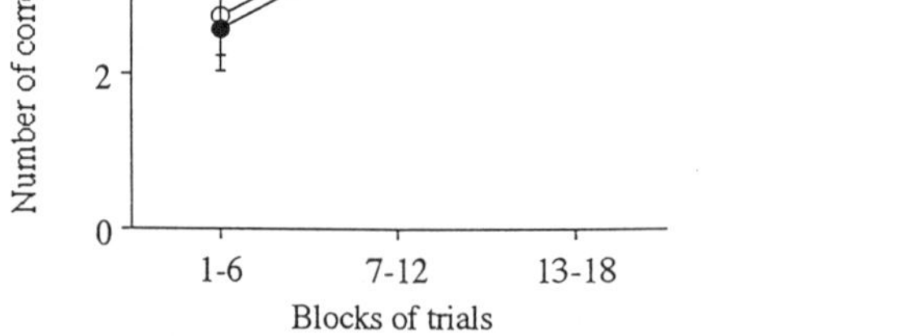

Fig. 7.3. Mean number of correct first choices (entering the baited table) in the first test trial over 6-day blocks for CNT, APC-ANT, APC-POST groups. Standard deviations are shown as vertical bars. (Reproduced from Thinus-Blanc et al., 1996b.)

The increasing level of exploratory activity that occurred across sessions in parietals seems to reflect the adoption of a compensatory strategy which improved performance. At the level of brain structure, such a compensatory strategy may be due to neural changes, particularly since structures such as frontal cortex or hippocampus can assume some functions normally mediated by the parietal area. Finally, unlike the Save et al. (1992b) experiment, APC-ant and APC-post rats displayed quite similar behavioral profiles.

5. The APC and Traditional Spatial Tasks

When tested in the Morris swimming task with a hidden platform, rats with APC lesions display a specific but mild impairment. They appear to locate the platform approximately, since during the probe trial with the platform removed they spend more time swimming in the quadrant where it was initially located (Kolb, Sutherland, and Whishaw, 1983). However, at the end of training, parietal rats continue to have difficulty in navigating to the exact location because of a persistent error in "heading" angles (i.e., the orientation of their initial trajectory is at a large angle from the actual direction of the platform). This error results in longer latencies, all the more so when the tank is of larger dimensions (Kolb and Walkey, 1987). Typical examples of swim paths of APC rats are shown on Fig. 7.4 (left). Similar deficits have been found in acquisition (Kolb et al., 1989) and retention (Whishaw, 1987) phases. With larger lesions, DiMattia and Kesner (1988b) found more pronounced deficits in both acquisition and retention.

The incapability of parietal rats to precisely locate the escape platform cannot be attributed to the fact that they have to swim, since in the "hole board" task that relies on the same principle as the Morris water maze task, similar data have been reported. After the rat has learned to locate the goal, a probe trial is conducted in which the food is transferred to a location diametrically opposite to the initial one. During training, rats with APC lesions have latencies and trajectories that are longer than those of the controls in reaching the baited hole. However, during the probe trial, they spend more time in the quadrant where the food was previously found (Kesner et al., 1989).

In another experiment in the Morris swimming pool by Kolb and Walkey (1987), a visual cue was placed within the tank (fixed on the wall), dissociated from the hidden platform whose location was changed from trial to trial ("landmark situation"). Spatial relationships between the cue and the platform remained the same across all trials. Extra-apparatus environmental cues were perceptible but irrelevant. The difficulty of localizing the platform in that case appears to be intermediately placed between the situation when the platform is visible, and when relevant information has to be picked up from the extra-pool environment. However, the APC-injured rats did not reach the platform as fast as controls. Interestingly, in this situation, animals displayed a specific "looping" strategy: they would typically swim along the wall of the pool, at a constant distance from it (Fig. 7.4, right). Unlike the trajectories observed in the traditional version of the task (without any intra-maze cue) which corresponded approximately to the direction of the platform, this strategy suggests that APC-injured rats were incapable to orienting, even roughly, toward the goal. They just knew that the platform was located a certain distance from the wall and swam until they found it. Kolb and Walkey (1987) proposed that the rats with

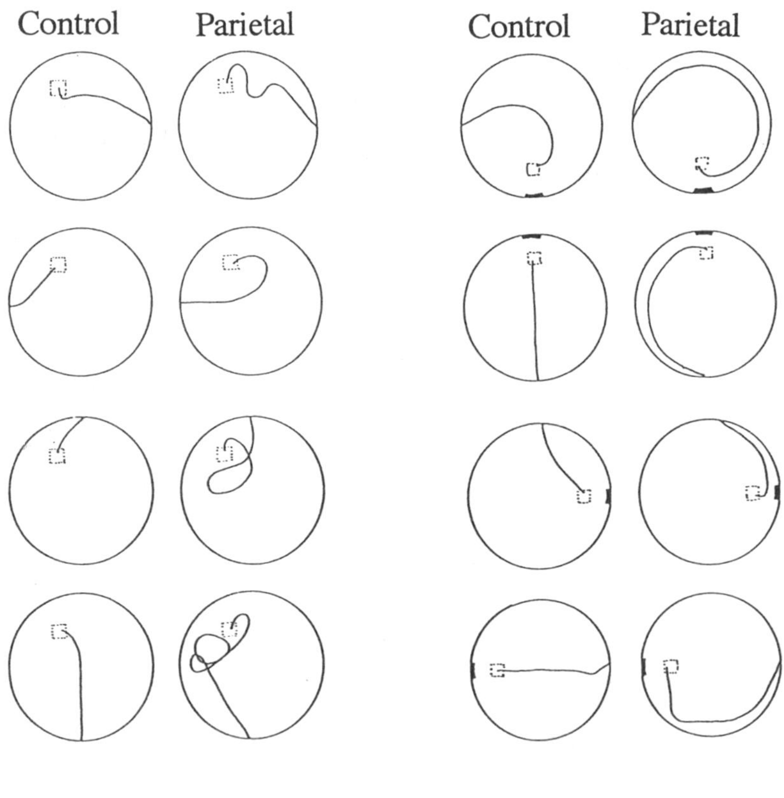

Fig. 7.4. Illustrations of the swim paths of a control and a parietal rat on the tenth trial block in the place navigation task (left) and in the landmark task (right). In both tasks, the control rat swims directly to the hidden platform whereas, in the place navigation task, the parietal is inaccurate in its heading, although this rat does navigate to the approximate location of the platform. In the landmark task, the parietal rat has adopted a looping strategy whereby it swims a fixed distance from the wall until it encounters the platform. (Adapted from Kolb and Walkey, 1987.)

parietal lesions were unable to extract the relevant cue (which was moved, with the platform, from trial to trial) from the larger set of irrelevant cues. However, the same authors found that when the platform was visible, moving it from trial to trial did not increase latencies in parietal rats. In that case, operated animals had no difficulty in disentangling the information provided by viewing the platform from other irrelevant environmental cues. Therefore, the deficit seems to be specifically related not only to difficulties in extracting the relevant cue from the set of irrelevant information but also in maintaining invariant the spatial relationship between the platform and the cue in spite of repeated discrepancies with the distal environment.

One hypothesis which is frequently proposed concerning the function of APC is that it ensures the integrative processing of visuospatial and movement-generated (kinesthetic) information. Consequently, in line with this hypothesis, whether reaching a goal requires egocentric or allocentric spatial organization, lesions of APC should impair orientation that is primarily based on kinesthetic information. To examine this possibility, Save and Moghaddam (1996) tested the role of APC in two experiments. In the first one, movement-related information was of crucial importance: APC-lesioned and control rats were trained to reach the submerged platform in a Morris swimming pool in total darkness. Under that condition, the relative positions of the starting point and the platform were constant across all training (egocentric task). Parietal rats have been found to be impaired in acquisition (Fig. 7.5) and, to a lesser extent, in retention of this task. It should be emphasized that even control rats had relatively long escape latencies

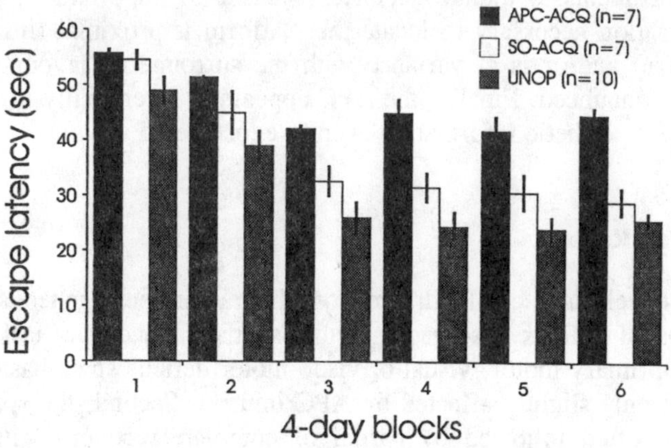

Fig. 7.5. Mean escape latencies in seconds during training (Days 1-24). Each bar represents a 4-day block. APC-ACQ: parietal-lesioned rats; SO-ACQ: sham operated rats; UNOP: unoperated rats. Note that even in SO-ACQ and UNOP groups, the latencies are very high; usually, they are around 10 sec. (Reproduced from Save and Moghaddam, 1996.)

at the end of training. When subjected to the traditional version of the navigation task (under normal lighting and with four starting points) the APC-lesioned rats displayed mild deficits in both acquisition and retention due to longer escape latencies, but with otherwise normal, control-like search patterns. Although APC-injured rats' trajectories were far more circuitous than those of the sham-operated or unoperated rats, their search was usually limited to the vicinity of the platform. Finally, in both conditions, heading errors decreased in all groups but remained significantly higher in the APC-group. The latter data (in the traditional version of the Morris task) confirm that parietal rats are not totally incapable of locating the platform, but this localization, as reflected by trajectory patterns, appears to be rather rough and imprecise. More importantly, the deficits observed in parietal rats on the egocentric task in total darkness demonstrate that the APC is involved in the processing of movement-generated information, the only information available in that situation.

The Olton radial arm maze is another apparatus which has been used to test the effects of APC lesions. All of the studies that have been conducted so far have revealed rather mild deficits in task acquisition, followed in most cases by near complete recovery (Kolb et al., 1983; Kolb and Walkey, 1987; Kesner, DiMattia, and Crutcher, 1987). Unfortunately in none of those studies was patterns of visits examined. Consequently, it is impossible to draw conclusions about the nature of compensatory mechanisms that permitted recovery.

In summary, APC-injured rats are capable of approximately locating a hidden goal in "open" situations such as the Morris swimming pool or the "hole board" task, with distal relevant cues, but they are impaired at efficiently making final adjustments to their trajectories because of important heading errors. When information necessary to locate the platform is proximal (inside the pool or the T-maze) and/or is at variance with the surrounding layout, deficits become more pronounced. Finally, the APC appears to be crucially involved in the processing of kinesthetic information even in egocentered tasks.

6. Concluding Remarks

Several conclusions can be drawn from this rather heterogeneous set of data. First, the spatial deficits observed in actual spatial (allocentric) tasks cannot be attributed to primary motor, visual or visuo-motor deficits since basic abilities at this level are only slightly affected by APC injuries. Second, the spatial deficits are mild and often followed by partial or complete recovery with additional training. Third, when observed, deleterious effects of APC damage apply equally to tasks which rely on egocentric or allocentric spatial organization. Finally, although parietal rats display quite normal patterns of exploratory habituation, they react accurately to a spatial change; nevertheless, this applies only when the change consists of displacing objects. Unlike control animals, they fail to search for a missing attractive stimulus by spending time at the place where it had formerly been located. However, the presence of a modified object configuration or the absence of an attractive stimulus-rat are not symmetrical situations (Foreman and Thinus-Blanc, 1987). The detection of a missing stimulus does not seem to involve the same mechanisms as orienting toward an unexpected distractor. The fact that in Foreman et al.'s (1992) experiment, both parietal and controls displayed a delayed reaction to the omission of a familiar distractor, whereas they reacted immediately to its unexpected occurrence, supports Foreman and Thinus-Blanc's previous conclusion. The lack of reaction to the removal of the stimulus-rat may be attributed to increased task difficulty by comparison with the situation where objects remain still present in the field.

The features of this "parietal syndrome" derived from lesion studies may lead us to conclude that the APC of the rat has a non-specific function that is far from being crucial for spatial processing. Such an assertion would be at odds with data derived from another approach, at the level of single-cell activity. Studies in this domain (plus a few results obtained from behavioral studies) have led McNaughton and his collaborators, among others, to the hypothesis that the APC is an interface between egocentric and allocentric encoding, which works by integrating kinesthetic and visual information while the subject is moving around in the environment. At the neuroanatomical level, the plausibility of this hypothesis is supported by reciprocal connections between the APC and the visual areas, on the one hand, and with vestibular nuclei on the other (see above). But neurophysiological data provide the strongest support for the hypothesis. In the radial maze, McNaughton et al. (1989b) have found cells in the parietal cortex that fire for specific combinations of environmental spatial features (local views) and specific movements. Recently, head-direction cells have been recorded in this area. These cells that are activated by the animal's head direction irrespective of location and behavior and are similar to those recorded from the hippocampus (see Chapter 6). The activity of some head-direction cells in the APC cortex have been found to be dependent on combined movements and visual or spatial features of the surroundings (Chen et al., 1994 a and b). According to these authors, this behavioral selectivity provides direct evidence that an integration of movement and directional (visual) information exists in spatial information processing. The APC seems to be highly involved in this integration, but it is also important in the use of purely movement-generated feed-back in the absence of visual information, as in the Save and Mogaddham (1996) experiment.

An integrative function of APC may also explain why deficits are more profound in parietals when relevant landmarks for locating a hidden goal are proximal than when they are provided distally by the extra-apparatus environment. Due to movement parallax, the visual motion through the visual field of a near stimulus, while the animal is moving around, is faster and more sensitive to small changes of direction than that of distal stimuli. This more "detailed" and precise input may increase the difficulty of integrating visual and movement-generated information. This could account for the moderately impaired accuracy of trajectories in the traditional version of the Morris swimming pool with distant landmarks, and for the more pronounced deficits with intra-apparatus landmarks. This hypothesis could be easily tested by varying the distance between the apparatus and the landmarks necessary for goal localization.

The integration impairment that emerges when APC-injured rats have to execute an oriented trajectory should also affect exploratory activity. In Chapter 2, I proposed that one of the functions of exploratory activity would be to enable the extraction of spatial invariants by matching visual and movement-generated information. Nevertheless, the habituation patterns of parietal-damaged rats have been found to be normal in all of the studies conducted so far. As suggested in Chapter 6, habituation rate may just reflect the fact that spatial information has been stored in memory. Unlike reactions to spatial changes, it does not say anything about the accuracy of spatial encoding. As a matter of fact, parietal-injured rats displayed either unselective re-investigation of displaced objects or a lack of reaction to a missing stimulus. There are two (non mutually exclusive) explanations for these deficits: First, the spatial encoding completed during habituation may have not been accurate enough for the spatial change to be precisely identified. This could account for the increased exploratory activity observed across days in the 3-table task. Unlike the situation in the short, unrepeated reaction-to-change tests, the whole duration of the 3-table experiment (18 days) would have allowed the animals to set up a strategy consisting of compensating for the poor quality of processed information by a quantitative increase.[a] Second, as in the Morris task, animals' lack of precision in adjusting trajectories toward the "goal" may in other circumstances, such as where they must move to displaced objects or to the initial location of a removed stimulus-rat, explain why no more contacts with displaced objects were made or time spent at the missing stimulus location. This means that besides the transformation of perceptual information into representations, the APC would achieve a similar transformation in the reverse direction, from representation to action, specifying, for instance, as motor commands, the directions to be taken, directions that have been previously computed at representation level.

---

[a] In this task the choice is binary and once it has been made, there is no need for adjusting the final trajectory in the alley. Consequently, no heading errors can be made as in the Morris swimming pool.

In conclusion, this overview of the literature leads to the conclusion that the APC plays a role in interfacing between egocentered and exocentered spatial processing. In this way, it would contribute to the transformation of information coming from various sensory channels (though mostly from vision), that is necessarily organized along the body referent, into allocentered representations that are independent of the subject's position. This transformation is achieved by matching the sensory input coming from the environment, with movement-generated feed-back. The latter provides the system with the necessary and crucial information that the animal itself is moving despite the permanently changing appearance of the surroundings perceived under different perspectives.

CHAPTER 8

THE SPATIAL FUNCTION OF THE PREFRONTAL AND OCCIPITAL CORTICES

"The apparent general similarity in prefrontal function across mammals may reflect the similarity in general requirements of mammalian behavioral patterns, whereas the differences in the details of prefrontal organization in mammals may reflect interspecies variability and the adaptedness of behavior." (Kolb, 1990b, p. 451)

Besides the associative parietal area, other neocortical regions, the prefrontal cortex and, to a lesser extent, the occipital cortex, appear to be involved in spatial processing. In spatial tasks, damage to the medial prefrontal cortex has been shown to result in an apparently heterogeneous pattern of effects, which vary according to task requirements. However, a good deal of data supports the idea that this region is involved in spatial behavior. The picture is far less clear concerning the occipital region of the neocortex classically considered to be exclusively responsible for the processing of visual features. However, some lesion data suggest that this region could also be involved in spatial processing, independently of the specific impairment of visual form perception. This idea is not new since Lashley (1943) and Tsang (1934, 1936) had found severe deficits in rats with destruction of the occipital area when they were tested on various spatial tasks. De Renzi (1982) in his book "Disorders of Space Exploration and Cognition" quotes several early clinical observations of patients with occipital damage who displayed orientational deficits (Foerster, 1890; Meyer, 1900; Peters, 1896), though these were casual observations the conclusions of which have not been investigated further

There are, of course, conceptual problems related to such a broad approach to the brain structures involved in spatial processing. One of them is to disentangle, in view of the deficits induced by neocortical lesions, what is specifically due to an impairment in spatial processing from what may be a consequence of a deficit at another level of processing. Sometimes, the boundary between spatial and non spatial impairments may be extremely subtle and difficult to define. That is the case with the occipital cortex. It is also possible that a given structure has a function at several levels, perhaps both spatial and non-spatial.

In the present chapter, the spatial functions of the prefrontal, and occipital cortices are examined. The effects of occipital lesions to occipital area on sensorimotor behaviors are described in order to examine the possibility that genuine spatial behavioral deficits may be a consequence of impairment occurring "upstream" in spatial processing.

1. The Prefrontal Cortex: Neuroanatomical Recall

A prefrontal cortex has been identified in the rat on the basis of connections between the medio-dorsal nucleus of the thalamus and prefrontal area which is the main target of medio-dorsal thalamic projections (Akert, 1964; Kolb, 1990b; Leonard, 1969). The prefrontal area has reciprocal connections with many subcortical and cortical regions, among them with the hippocampus and parietal cortex (Divac, Kosmal, Björklund, and Lindvall, 1978; Reep, Goodwin, and Corwin, 1990; Uylings and van Eden, 1990). The rat's prefrontal cortex is subdivided into two main regions: the medial prefrontal and orbital (or sulcal) cortices (Divac et al., 1978; Groenewegen, 1988; Krettek and Price, 1977; Uylings and van Eden, 1990). These two regions are themselves made up of various subregions that are thought to be functionally distinct.

Recently connections have been demonstrated between one of the subregions of the median prefrontal cortex, the prelimbic area, and various subcortical structures such as the central grey nuclei (Condé, Audinat, Maire-Lepoivre, and Crépel, 1990; Gronewegen and Berender, 1994) suggesting that prelimbic cortex could be heavily involved in the execution of action. Moreover, there are direct functional connections from the temporal hippocampus to the prelimbic area of the rat (Jay, Glowinsky, and Thierry, 1989; Jay and Witter, 1991; Laroche, Jay, and Thierry, 1990; Swanson, 1981). Unfortunately, no studies have been conducted so far to compare the behavioral effects of discrete lesions in subregions of prefrontal cortex.

2. Effects of Prefrontal Lesions on Reaction-to-Change and Spatial Problem Solving Tests

Poucet (1989) tested rats with medial prefrontal lesions, examining their reaction to a spatial change following habituation in an open-field containing four objects. Frontal animals were similar to normal subjects with respect to their initial exploratory level and habituation pattern. In addition, they re-explored the displaced object only, and completely neglected the nondisplaced ones. Thus, they reacted selectively and more accurately to the change than controls which, in this situation, reinvestigated all of the objects.

In a further study, Granon, Save, Buhot, and Poucet (1996) tested frontal rats' behavior in two tasks requiring more effortful processing. The lesions were restricted to the prelimbic area. First, the experimenters exchanged two objects for the test (instead of displacing only one of them). Thus, unlike Poucet's experiment, the geometrical arrangement remained the same but only vicinity relationships among objects were modified. In addition, the authors compared the effects of frontal lesions, as related to the number of items that had to be processed. In one of the situations, only three objects were put into the field whereas the other situation included six objects. The hypothesis was that if frontal rats have a specific problem in effortful spatial situations they would be less likely to detect a spatial change occurring in a more complex environment (two exchanged objects out of six) than a spatial change occurring in a simpler environment (two exchanged objects out of three).

As shown previously, frontal rats displayed normal exploratory activity and habituation. Moreover, in both normal and frontal rats, the renewal of exploration following the change was specifically directed toward the displaced objects. This reaction was observed in both conditions, a result which indicates that frontal lesions did not preclude animals from detecting the spatial change, even in the complex situation with six objects.

Therefore, whether frontal lesions are restricted to the prelimbic area or entirely damage the medial prefrontal cortex, they have no deleterious effect on the initial level and time-course of exploratory activity, nor on reactions to spatial changes.

The Maier-three-table spatial reasoning task (see Chapter 2) has also been used to test the effects of frontal cortex lesions. Herrmann, Poucet, and Ellen (1985b) showed that rats with medial frontal lesions are unable to perform as successfully as normal rats. Their performance did not improve with repeated testing. Instead, the test-trial performance of medial frontal animals was characterized by the repeated use of the same directional response at the choice point (i.e., the adoption of a body-turn strategy.) Normal rats habituated, i.e., decreased the number of table entries within the first few minutes of the daily exploratory period. Although medial frontal rats initially failed to display this habituation pattern, they did develop habituation over blocks of trials.

In another experiment, already described in Chapter 6, Poucet and Herrmann (1990) used two experimental situations aimed at decreasing the difficulty of the spatial reasoning task, one with a fixed-goal (FG) situation (the start table was randomly varied from day to day), another with a variable-goal (VG) situation (the start table remained constant). In both tasks, normal animals performed better than frontal rats whose performance, nevertheless, did improve over days, though in the FG only, and not in the VG task. This pattern of results is similar to that observed following septal lesions (Fig. 6.9). As found previously, frontal rats did not demonstrate abnormal exploratory activity.

On the basis of these data, Poucet (1990) attempted, first, to alleviate still further the deficit in performance displayed by frontal rats in the VG task by adding visual cues either on each table and its associated runway or on the goal table (Experiment 1). Second, he examined the hypothesis that the gradual improvement of frontal animals in the FG task in the previous study resulted form the use of a compensatory strategy simpler than building an overall representation (such as associating a certain cue with the goal, i.e., implementing a "guidance" according to O'Keefe and Nadel's [1978] terminology). In order to prevent the use of cues directly associated with the goal location, several maze positions were used within the experimental room (Fig. 8.1) and animals were required to orient from various starting points (and consequently under different perspectives, involving different landmarks) to a single goal location common to all maze positions (Experiment 2). The results of Experiment 1 showed that the addition of distinctive visual cues on the tables and their associated runways, or

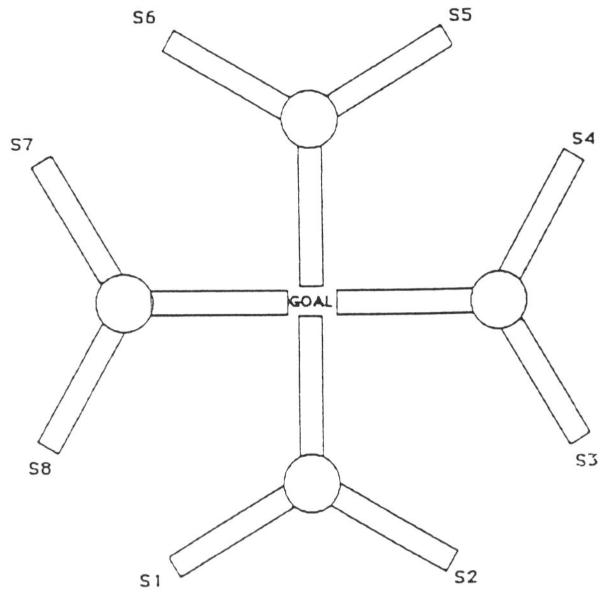

Fig. 8.1. A schematic representation of the 4 possible positions of the three-table apparatus used in Experiment 2. The corresponding start locations are indicated by S1, S2, ... S8. The tables (not drawn for sake of clarity) were disposed at the extremity of each of the three arms of the apparatus. Whatever the actual position of the maze in space, the goal table was consistently located at the center of the experimental room. (Reproduced from Poucet, 1990.)

a conspicuous visual pattern placed directly above the goal, failed to improve frontal rats' performance. In contrast, the daily displacement of the apparatus in order that animals could not learn to associate the goal table with specific cues located behind it did not prevent frontal rats from learning the consistent location of the food by using, probably, its relationship to the configuration of landmarks provided by the experimental room.

In summary, as during the exploratory phase prior to reaction-to-change tests, habituation is not impaired in the three-table maze. Deficits are observed in frontal rats in the traditional version of the task when the goal location is changed each day, i.e., when short-term Working Memory (WM) is involved, despite the fact that additional information may have allowed animals to implement simple guidance strategies. In contrast, when the goal is at a fixed location over days, no impairment is observed in frontal animals even though the difficulty of the task is increased by a large number of starting points. All together, these data demonstrate that prefrontal lesions do not preclude (1) the encoding of spatial information, and (2) the efficient use of spatial information stored in long-term memory. In contrast, if short-term storage of the goal location is required, deficits are consistently observed in frontal animals.

3. The Prefrontal Cortex and Traditional Learning Tasks

According to the data presented above, rats with prefrontal lesions should display WM deficits in the eight-arm Olton maze with four arms baited, i.e., errors should be due more to re-entries of initially baited arms (termed WM errors) than to visits to arms that have never been baited (termed RM errors; see Chapter 3, for a discussion of this distinction). Indeed, frontal lesions were found to result in impaired performance in the radial maze in virtually all the studies which have been conducted so far. Unexpectedly, the errors made by the medial-frontal-lesioned rats were in learning to discriminate rewarded from non-rewarded arms (RM errors) and not in discriminating arms they had just entered from arms they had not yet entered (WM errors; Kolb, Pittman, Sutherland, and Whishaw, 1982; Kolb et al., 1983; Becker, Walker, and Olton, 1980). Similarly, Kesner et al. (1987) found that rats with medial prefrontal cortex lesions displayed deficits in both WM and RM followed by complete recovery of WM and a partial recovery of RM.

Deficits displayed by prefrontal rats in the Morris swimming-pool (Kolb et al., 1982) are apparently indistinguishable from those observed following hippocampal (Sutherland, Kolb, and Whishaw, 1982) or posterior parietal damage (Kolb et al., 1983). Four starting points are usually used as early as the first day of training. Frontal rats have been found to need much more time to reach the escape platform but, after an extended period of training, and despite persistently large heading errors, their latencies fell to levels no worse than those of controls. The authors concluded that such data provide evidence of a significant spatial orientation deficit in rats with prefrontal lesions.

However there are a variety of stages and processes that are included under the generic term "spatial deficits". For instance, deficits can be due to an impairment while acquiring spatial knowledge, or while using this knowledge for orienting. In a recent experiment using the water maze, Granon and Poucet (1995) have provided additional support for the idea that the observed spatial deficits in the Morris swimming pool are due to a WM dysfunction that would preclude animals from forming an adequate representation of the whole course of movements required to reach the platform. Fig. 8.2 is a diagrammatic representation of the schedule that was used. Stage 1 consisted of 5 consecutive sessions during which the rats were trained to reach the hidden platform location from a single starting position that was used on all trials. During stage 2, rats were required to reach the same platform location from a new start position (SP2) as well as from the previously experienced start position SP1. During stage 3, the platform was relocated in the opposite quadrant. The rats were then tested with the same two start positions as during stage 2. Stage 4 involved the same platform location as Stage 3, but 4 distinct start positions were used. They were the two used in Stages 2 and 3 plus two new starting positions, SP3 and SP4.

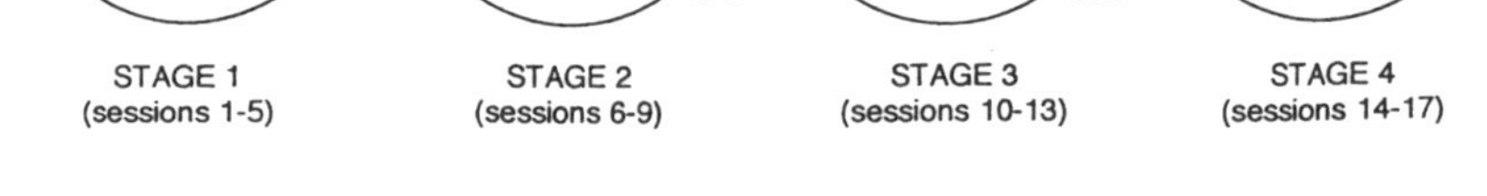

Fig. 8.2. A diagrammatic representation of the experimental schedule. The goal platform location is indicated by a solid black square. The former location of the platform is shown as a white square delineated by hatched lines. Each session consisted of four trials. SP = start position. (Adapted from Granon and Poucet, 1995.)

In addition to the recording of latencies and distances, the cumulative distance to the goal was calculated on each trial (see Gallagher, Burwell, and Burchinal, 1993; see discussion in Granon and Poucet, 1995). Briefly, the distance between the current position of the rat and the goal position was calculated for each even sample (separated from each other by 160 ms). Adding these distances provides a measure of the *cumulative search error* which reflects deviation from an optimal path (Fig. 8.3). Another measure, *search error dispersion*, was intended to reflect the dispersion of searching behavior independently from the distance swum by the rat. For that purpose, the cumulative search error was weighted by the total number of even samples (for details, see Granon and Poucet, 1995; and Benhamou, 1989).

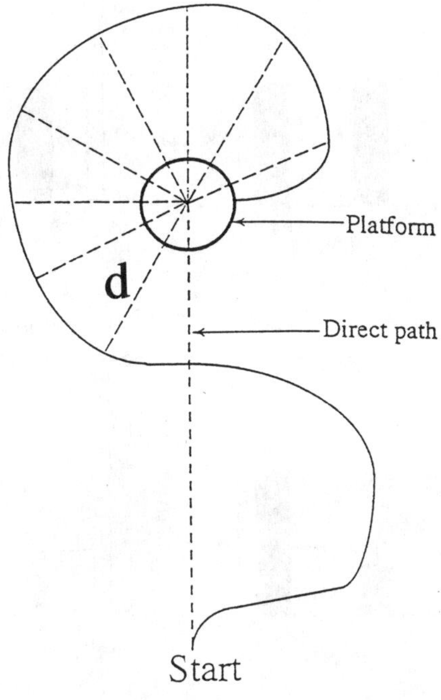

Fig. 8.3. Example of the computation of the cumulative search error. While the rat is swimming, the distance "d" between the current position of the rat and the goal is sampled every 160 ms. This sampling is made on the length of the total path less the distance corresponding to a direct trajectory. Adding these values provides a measure of the cumulative search error. The search error dispersion is computed by weighting the cumulative search error by the total number of even samples.

During the first three stages, rats with prefrontal lesions had performance levels (latencies, distances, cumulative search error and search error dispersion) similar to that of the controls and they accurately adapted their trajectories to both the new starting position and new platform location. It was only during Stage 4, with four distinct starting positions that a deficit appeared (Fig. 8.4). It was specific to the two start positions newly introduced during this stage. Unlike latencies, the values of cumulative search error and search error dispersion emphasize frontal rats' impairment with the new starting positions.

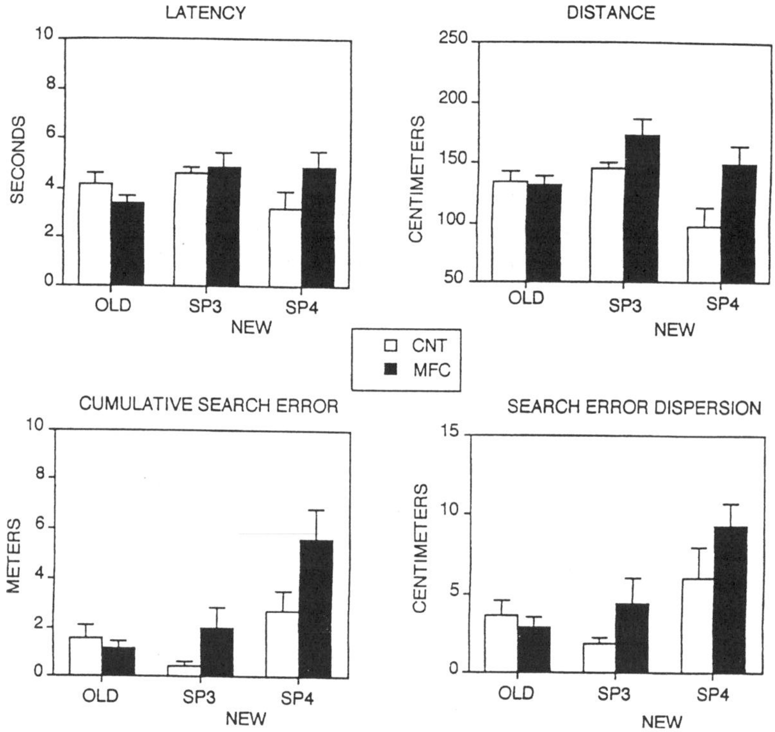

Fig. 8.4. Navigation scores from each of the two "new" start positions, SP3 and SP4, during stage 4. Values are expressed as means plus or minus standard error. For comparison, the mean navigation scores obtained from the "old" start positions are also shown. CNT = sham-operated group; MFC = medial frontal cortex. (Reproduced from Granon and Poucet, 1995.)

The data obtained using the Morris swimming pool allow us to identify the functional role of the prefrontal cortex. Together with normal habituation and reaction to a spatial change, the observation of normal navigation in frontal rats during the first three stages of Granon and Poucet's study supports the notion that prefrontal rats are *not* impaired in coding spatial information. In contrast, deficits appear when spatial knowledge has to be converted in oriented behaviors, and that, under certain conditions only.

Generally, rats with frontal lesions are impaired at many delay-type tasks in which an interval is imposed between stimulus presentation and the corresponding associated behavior. Such tasks include the three-table spatial reasoning test, but also delayed alternation (Divac, Wikmark, and Gade, 1975; Larsen and Divac, 1978; van Haaren, de Bruin, Heinsbroek, and van de Poll, 1985), and delayed matching and non-matching to sample (Brito and Brito, 1990; Brito, Thomas, Davis, and Gingold, 1982; Kolb et al., 1989). We have recently demonstrated that the frontal lesion-induced WM impairment is larger in spatial matching- than in non-matching-to-sample when the sample is a spatial location and the response is a spatial response (Granon, Vidal, Thinus-Blanc, Changeux, and Poucet, 1994). All of these tasks unambiguously involve WM, i.e., the short-term storage of information which has to be used subsequently for executing an action. The fact that frontal animals are impaired at processing temporal information (Kesner, 1989) is in agreement with this interpretation.

The lesion evidence is complemented by a few electrophysiological studies that have reported increased firing of single units in the frontal cortex during the delay period of delay-type tasks (e.g., Batuev, Kursina, and Shutov, 1990; Sakurai and Sugimoto, 1986).

At first glance, navigating in the Morris swimming pool does not require WM since the information necessary to reach the goal is constantly available. However, observations of intact rats subjected to various lighting conditions (see Chapter 3, section 7) indicate that visual perception of the environment is crucial at the beginning of the trajectory, while swimming a certain initial distance from the starting point (calculated to be about 30 cm in Save's [1996] experiment). After this, visual information is not necessary in order for the animal to reach the escape platform. This means that, under those particular lighting conditions, the final part of the trajectory has already been planned when visual information was still available. Even when testing is conducted with permanent lighting, there is no compelling need for planning trajectories, yet it is a fact that this occurs: straight, fast swimming without any hesitations is routinely observed after a few days of training. This planning process clearly involves WM, but less so in the swimming pool than in the three-table task. In the latter situation, animals are removed from the apparatus after having been shown a baited table whose location changes each day, and prior to undergoing the choice test. Consequently, they have to learn to pick up information relative to the goal location relatively quickly, and long before it becomes necessary.

In view of the differences between the three-table and the Morris navigation tasks, it is not surprising that frontal animals' deficits are far more pronounced in the former than in the latter. In the swimming pool, they are capable of reaching the hidden platform, as in Granon and Poucet's experiment (1995) so long as the situation is not very complex (in that case, when a small number of starting locations are used). Conversely, acquisition is impaired in frontal rats when they are tested, right from the first training day, using the traditional procedure with four starting positions, as in the experiments by Kolb and collaborators. The fact that frontal rats are capable of reaching the goal in a constant location from a number of starting locations in the three-table tasks (Poucet, 1990, see above) but not in the Morris navigation task may appear contradictory. Indeed, in an actual maze such as the Maier three-table situation, the alley configuration allows only a limited number of choices, which reduces the difficulty of the task. In those cases when the goal is at a fixed location (Poucet, 1990), this facilitation may compensate for the difficulty related to the delay between the presentation of relevant information (which does not change from one day to the next) and the choice. In contrast, in the Morris swimming pool, though relevant information is available all the time (under constant lighting conditions), the number of degrees of freedom is virtually infinite and trajectories have to be precisely planned if the animal is to escape from the water as fast as possible.

According to the terminology currently in use (see Chapter 3 for a discussion of this issue) RM deficits observed in the radial maze appear to be more difficult to account for within the above explanatory framework. These deficits seem to indicate a true spatial impairment in discriminating between the two subsets of arms (baited and unbaited). However, long-term storage in RM of the invariant features of the situation (in that case, that there are two subsets of arms) is acquired on the basis of short-term storage, within each trial, of the outcomes of the visits. Thus, one might expect that prefrontal rats would make as many RM errors as WM because they would fail to learn to discriminate between baited and unbaited arms. RM errors would be the consequence of a WM deficit which should also appear within a trial as an increased number of revisits to emptied arms. As a matter of fact, past authors have observed more RM than WM errors. Kesner et al. (1987) found a complete recovery of prefrontal rats' WM deficits. The only explanation of this puzzling result, which stands in contrast with all of those obtained in other experimental situations, is that frontal rats have implemented compensatory strategies after many days of training. In this respect, it would be appropriate to carefully examine whether the patterns of visits made to the baited arms after complete recovery could be due to the adoption of some form of stereotypical choice strategy or pattern.

Some authors have also suggested a crucial role for rat prefrontal cortex in tasks that require high attentional control (Brown, Bowman, and Robbins, 1991; Olton, Wenk, Church, and Meck, 1988) such as navigating in the radial arm maze or in the Morris swimming pool. However, Foreman (1983) showed that overt orienting responses to distracting stimuli are as normal in prefrontal rats as in controls. Furthermore, in Granon et al.'s (1996) experiment (see above), frontal lesions did not impair reactions to a spatial change though the situation was rather complex. These data run against the hypothesis that a primary attentional deficit occurs in frontal rats, whether it is related to the capability to shift attention to unexpected stimuli or to focus it for effortful processing.

4. The Occipital Cortex: Anatomical Recall

Although the precise number and locations of distinct visual areas in the rat is a controversial issue, it is broadly acknowledged that the occipital cortex is made up of three main regions: the striate cortex (Oc1 or area 17), and the lateral (Oc2L or area 18a) and medial (Oc2M or area 18b) extrastriate cortices (Dean, 1990). With respect to spatial processing, it is noteworthy that visual areas have direct and indirect connections with a large number of cortical regions such as parietal area 7. Visual information reaches the hippocampus through two main pathways, one of which is subcortical via thalamic nuclei; the other is neocortical through the entorhinal cortex. An important aspect is that, in rats, unlike primates, the dorsal lateral geniculate nucleus of the thalamus receives a relatively small proportion of retinal input in comparison with the

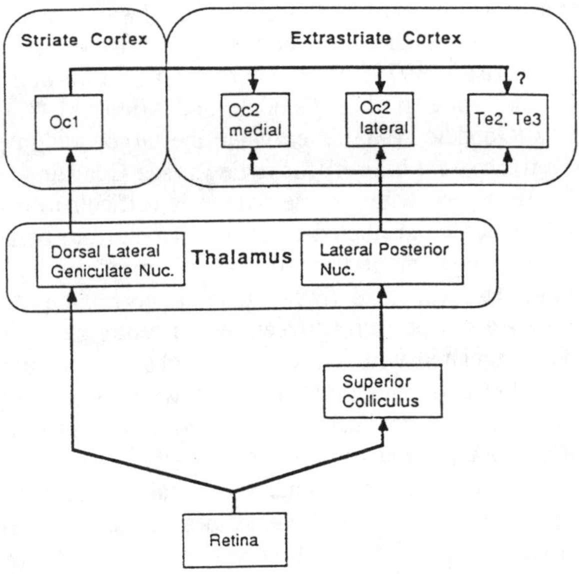

Fig. 8.5. Simplified schematic diagram of the rat's visual cortex to illustrate its major visual input. Efferent projections of extra-striate areas are not shown, nor are subcortical projections of Oc1. (Reproduced from Dean, 1990.)

superior colliculus which is part of the visual system (Fig. 8.5). All the retinal ganglion cells project to this midbrain structure, and less than 40% of them have collateral projections to the dorsal lateral geniculate nucleus (see Goodale and Carey, 1990). In functional terms, that means that the superior colliculus which is an important visual and sensorimotor structure in rodents (see Thinus-Blanc, 1983, for a review; Thinus-Blanc and Lecas, 1985; Scardigli, Fabre, and Thinus-Blanc, 1990; Thinus-Blanc, Scardigli, and Buhot, 1991), could be involved in compensatory mechanisms after lesion to the occipital cortex.

Moreover, Toldi, Fehér, and Wolff (1986) have discovered regions of multimodal integration around primary visual areas. These regions contain neurons that respond to each of the modalities specifically processed by the neighboring auditory and somatosensory cortices. This finding has important implications as to the integrative role of vision (see section 8).

5. Effects of Occipital Lesions in Visuo-Motor Tests

Besides studies aimed at examining the effects of occipital lesions on specific visual tasks (for a review, see Dean, 1990), many others have been concerned with the effects of such lesions on sensorimotor behaviors.

The performance of rats with visual cortical lesions is severely impaired in visually guided tasks such as locomotion toward low-contrast target door (Goodale and Murison, 1975) or to a target hole in an arena covered with a vertical black and white grating (Goodale and Milner, 1982), and locomoting around barriers (Goodale, 1983) even when the target was in plain view on the other side of a transparent barrier (for reviews, see Goodale, 1983; Goodale and Carey, 1990). However, within these data it is difficult to disentangle to what extent the deficits can be attributed to an actual failure to perceive the target as distinct from desembedding it from the visual array, or to a memory deficit inducing a failure to recognize (rather than perceive) the target. Recognizing involves making a a comparison between the currently perceived stimulus and a stored stimulus associated with a reward. An observation by Schneider (1966, 1969) illustrates this differentiation. Hamsters with lesions of posterior neocortex were unable to perform horizontal-vertical discriminations but they could orient as normal animals toward sunflower seeds presented anywhere in their visual field. Since they were able to perceive small seeds, it is likely that, a fortiori, they could also perceive the large stripes of the discriminanda. However, in the latter case, the cognitive requirements of the task were higher since, in addition to the perception of the patterns, they had to remember which kind of stripe was associated with the reward in order to perform the discrimination. This result is consistent with the fact that lesions of primary visual cortex only slightly impair the estimation of distances when gerbils were required to jump from one platform to another (Carey, Booth, and Goodale, 1987). In contrast, Lashley showed in the rat (1939) that large lesions in the occipital region induced a deficit in a learning-set task (consisting of choosing the bigger or the smaller of two stimuli in a traditional discrimination paradigm). Therefore, a possible conclusion is that the lesions had affected the visual processing of "artificial" complex stimuli, leaving intact the perception of simple "natural" objects and distances which might have been mediated by other brain structures such as the superior colliculus (the so-called "secondary" visual system). A similar conclusion is drawn by Goodale and Carey (1990) about the discrepant data obtained in distance estimation and discrimination tasks: in the latter case, the artificial association of a value with a stimulus that is otherwise meaningless for the animal is much more "cognitive" and may therefore depend on quite plastic cortical mechanisms.

6. Effects of Occipital Lesions on Reaction-to-Change and Spatial Problem Solving Tests

A non visual version of the dishabituation experiment has been recently designed (Krimm, 1994; Thinus-Blanc and Krimm, 1993) to test rats with occipital damage (all of the lesions including area 17, and most of them areas 18a and b also). The aim was to determine whether non-visual spatial processing is affected by occipital injury. For that purpose, instead of an open field containing objects, a cross-maze shaped apparatus was used, placed in a curtained visually homogeneous environment. The arms were visually identical but differed from each other either by having flower scents wiped on the distal inside part of the wall or by a buzzer fixed outside the distal wall (Fig. 8.6). A covered central area served as a shelter where animals stayed either before starting to explore or once they had completed exploration. The time spent in each of the arm was recorded. After three phases of exploration, two of the arms were interchanged and the reaction to this spatial change was recorded. A sham-operated group of rats was split into two control subgroups. One of them was subjected to the spatial change, whereas the situation remained unchanged for the second one in order to check that the expected increase in time spent in the alleys following the change was actually due to the latter and not to some uncontrolled factor. The data shown in Fig. 8.7 show that in spite of a higher level of activity during the first two phases of exploration, occipital rats displayed normal habituation and reacted to the spatial change by spending more time in the alleys than during the last session of habituation. Similar data were obtained in a visual version of this reaction-to-change test (unpublished study).

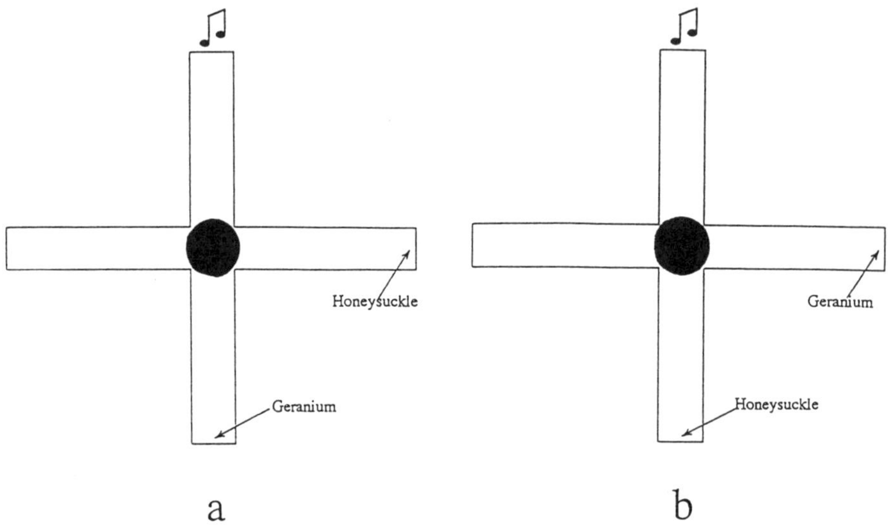

Fig. 8.6. Schematic representation of the cross-shaped apparatus to test reactions to a spatial non-visual change during (a) habituation and (b) test. The black circular area represents the covered shelter. (Adapted from Krimm, 1994.)

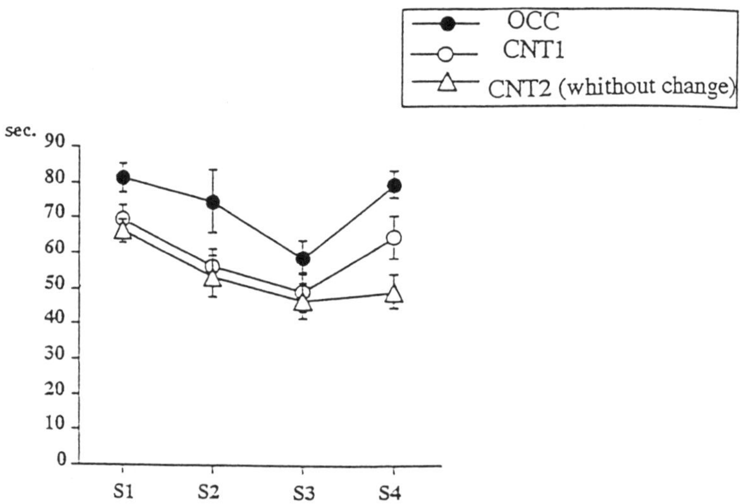

Fig. 8.7. Evolution of exploratory activity (time spent in the arms of the maze) during habituation (S1-S3) and during the test (S4) for the occipital lesioned (OCC), control with change (CNT1) and control without change (CNT2) groups. (Adapted from Krimm, 1994.)

The same operated rats were also tested on the Maier three-table test (Krimm, 1994). The patterns of habituation of visited tables during each daily exploratory phase were found to be roughly the same as those observed in rats with posterior parietal lesions. They did not differ from the controls during the first six-day block but the level of activity increased (with intra-session habituation) during the last two blocks in the same way as it has been observed in parietal rats (Chapter 7). Although their performance level on the tests increased at block 2, unlike parietal rats it did not reach that of the controls at the end of the experiment (Fig. 8.8).

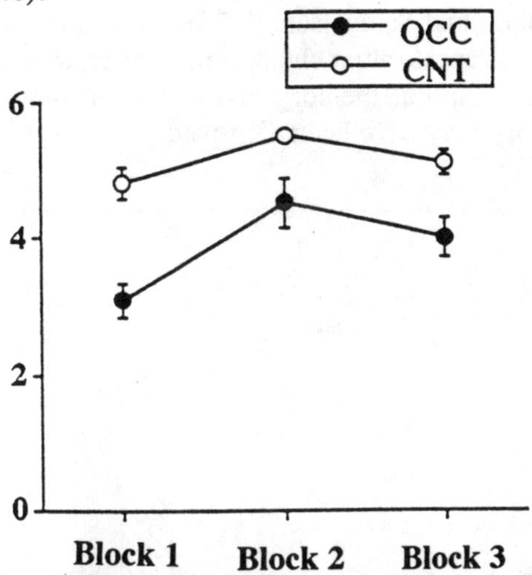

Fig. 8.8. Mean (±S.E.M.) Trial 1 performance (% correct) over 6-day blocks for the occipital-lesioned (OCC) and sham-operated (CNT) groups. (Adapted from Krimm, 1994.)

7. The Occipital Cortex and Traditional Learning Tasks

The most systematic studies in animals have been conducted in rats using the radial arm maze (see Chapter 3). The role of visual cues in this task has been demonstrated by several authors (e.g., Olton and Collison, 1979; Zoladek and Roberts, 1978; Suzuki et al., 1980). Therefore, it is not surprising that blinded rats need more trials to reach the learning criterion than sighted ones (Dale and Innis, 1980). This is also the case with animals trained in darkness (Dale, 1980). However, enucleation after reaching the learning criterion slightly impairs the performances in the radial arm maze (Dale and Innis, 1980), whereas occipital cortical lesions induce substantial deficits (Foreman and Stevens, 1982).

The same striking differentiation between peripheral enucleation and central intervention on the visual system has been found by Goodale and Dale (1981).

After reaching the learning criterion in the eight arm radial maze, some of the rats were deprived of vision by enucleation, others sustained lesions of the occipital cortex. A third group received both kinds of intervention. A fourth group was sham-operated. Again, the rats blinded by enucleation recovered a good level of performance. In contrast, those with lesions of the visual cortex, regardless of whether an enucleation had been carried out, displayed dramatically impaired performance which did not improve with time (Fig. 8.9). In addition, the intact and enucleated rats showed the well-known tendency to repeat their choices (i.e., to make errors) on the arms which had been visited first, at the beginning of a trial (Olton and Samuelson, 1976). In contrast, all the animals with cortical lesions did not repeat initial choices more often than later ones. This last finding suggests that temporal memory and spatial memory as well as spatial components of memory may have been disrupted.

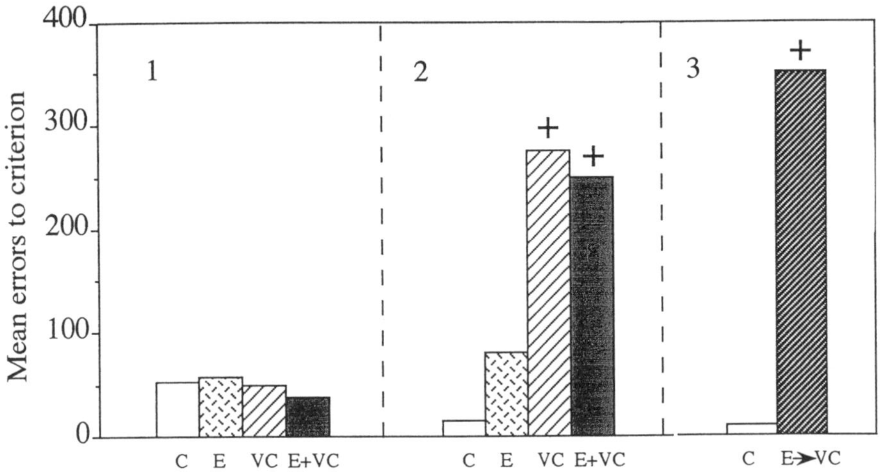

Fig. 8.9. Total number of errors required to reach criterion in each of the three phases of the experiment. Group C received only sham operations; Group E, enucleations and then, Group E → VC, cortical lesions; Group VC, cortical lesions; and Group E + VC, enucleations and cortical lesions. (Adapted from Goodale and Dale, 1981.)

Most interestingly, similar cortical lesions made after reaching criterion in animals which have been enucleated before learning did not disrupt the performance in the radial arm maze (Goodale, 1983).

These data are all the more significant because, unlike peripheral enucleation-induced blindness, visual cortical removals do not totally abolish visual perception in rats (Ferrier and Cooper, 1976; Goldstein and Oakley, 1987; Mize,

Wetzel, and Thompson, 1971). The "secondary" visual system, i.e., the extrageniculostriate system, and particularly the retinotectal pathway, is relatively well developed in rodents and known to subserve coarse visual processing (Dean and Redgrave, 1984). This implies that even with damage to the visual cortex, rats should still be able to use gross visual cues relevant in the radial maze task. In contrast, peripherally blinded rats must have relied exclusively on non-visual strategies. Therefore, the observed deficits do not appear to be dependent on the amount of available visual information but on a particular role of the visual cortical areas.

In the water maze, impaired navigation has also been observed in occipital rats in both acquisition and retention ((Krimm, 1994; Thinus-Blanc and Krimm, 1993). The pool was placed within a curtained environment with a single conspicuous large black card fixed to the inside wall. This procedure was used in order to ascertain that occipital rats were capable of perceiving the visual cue necessary to locate the hidden platform (visual tests had been conducted previously). The procedure was otherwise traditional, with a probe trial being conducted without the platform visible. Although rats with occipital injury did learn the task, their latencies were longer than those of the controls at the end of training (Fig. 8.10a). However, this impairment was rather mild and these animals spent more time, during the probe trial, in the quadrant where the escape platform was located. In contrast, occipital damage was found to have a more pronounced effect in retention (Fig. 8.10b). Latencies were longer and did not decrease across sessions; no more time was spent in the quadrant where the platform was initially located. The search error dispersion index (see section 3 above) calculated for pre- and post-learning operated rats indicated that occipital lesions had a more deleterious effect in the latter group where many erratic search trajectories were observed.

A puzzling result deserves to be mentioned. At the end of training, for all the groups, an additional test-trial was conducted with no cue-card inside the pool.[a] That was an additional way of checking the capability of the occipital rats to visually perceive the landmark, supplementing the simple visual discrimination test had been conducted before the experiment with a black and a white card. Unexpectedly, removing the only available visual cue had a deleterious effect in none of the groups. It is possible that rats had learned to use distal non visual cues. Concerning the occipital rats, it allows us to rule out emphatically the possibility that erratic search patterns may have been related to a pure visual deficit.

---

[a] This test occurred after the probe trial without platform, but a session in normal conditions was conducted meanwhile in order to assess that animals had not been disturbed by experiencing the pool without escape platform.

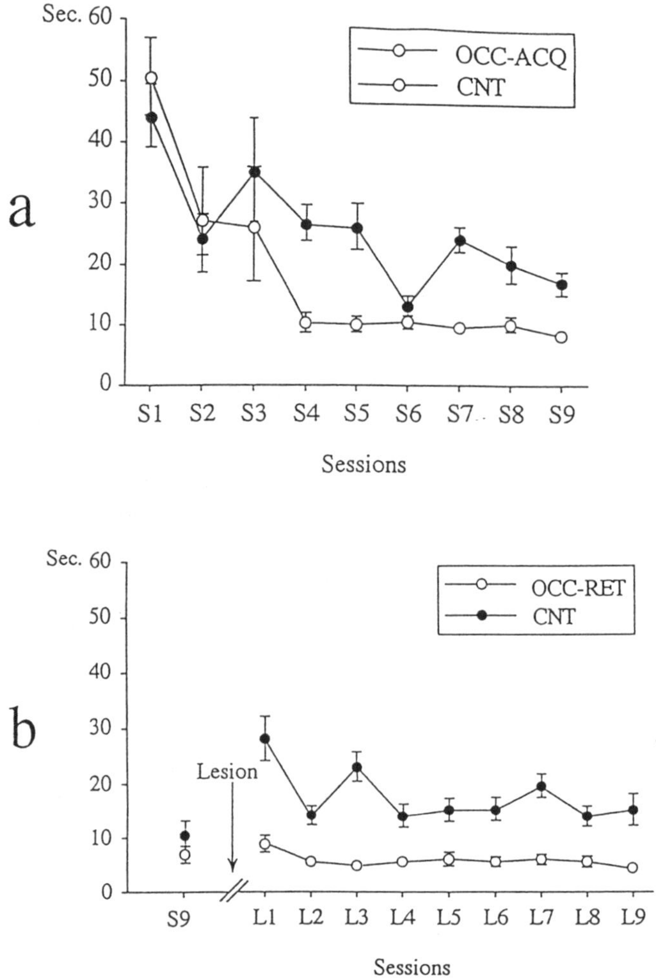

Fig. 8.10. Mean latencies to reach the hidden platform during (a) acquisition (OCC-ACQ: occipital lesioned rats; CNT: sham-operated rats) and (b) retention (OCC-RET: occipital-lesioned rats; CNT: sham-operated rats) in the Morris swimming pool. (Adapted from Krimm, 1994.)

In summary, the data obtained in traditional learning situations indicate that rats with occipital damage do have spatial deficits that are not a consequence of a visual impairment. The impairment is rather mild when surgery is made prior to learning in the Morris swimming pool, the only situation where the effects of occipital lesions on acquisition have been studied. The fact that occipital rats' performance was found to be more affected during retention than acquisition is in agreement with all of the results that have been obtained in the radial maze (see above), with rats operated after reaching the criterion with visual cues available. In contrast, where vision has been made unavailable during acquisition, occipital lesions do not disrupt performance (Goodale, 1983). Consequently, although the occipital cortex contributes to the acquisition of a spatial task, the few data available so far show that operated animals are capable of improving their performance although they fail to achieve the performance level of controls. However, once the occipital region has been involved in learning, more pronounced deficits are then observed.

An idea that was developed in Chapter 5 (see also Chapter 10) is that an important element for spatial navigation is place recognition, which implies that visual scenes (or local views) as seen from specific places have been stored in memory. Stored local views can be placed in the same category as humans' mental images although the term is never used in animals.

In human studies, it is becoming gradually accepted that the representation of a visual scene or mental image activates the same brain structures – located in the occipital region – as those involved in the perception of this scene. In accord with this hypothesis, an explanation of the occipital-injured rats' marked deficits in retention is that damage to this region would destroy the site of storage of the local views used to solve the task during acquisition. The next section provides a brief overview of human studies concerning the neural substrate of mental imagery.

8. Occipital Cortex as the Neural Substrate of Mental Imagery and Storage of Local Views

In his book "Image et Cognition", Denis (1989) provides an extensive review of the studies supporting the notion of a common neural substrate for visual perception and imagery. This notion has been developed by many psychologists such as Neisser (1967), Shepard (1975; Shepard and Podgorny, 1978), Finke (1980, 1985), and Kosslyn (1980; Kosslin and Schwartz, 1977). At the level of brain mechanisms, various approaches provide convergent data. From the detailed analysis of a large number of brain-damaged subjects, Farah (1984) concluded that deficits which could be attributed to a loss of the image generation components were correlated with damage in the posterior left quadrant of the brain. Similarly, in all but three of the 13 cases of loss of long-term visual memories examined by Farah, the damage involved one or both occipital lobes.

Convergent data demonstrating that visual images and visual percepts share the same neural substrates have been obtained with other methods. By recording event-related potentials to stimuli from scalp electrodes in normal subjects,

Farah, Péronnet, Gonon, and Giard (1988) demonstrated that an effect of imagery was observable, localized directly over visual cortex. The analysis of the patterns of regional cerebral flow, measured during visual imagery tasks, has led to similar conclusions (Goldenberg, Podreka, Steiner, and Willmes, 1987; Goldenberg, Podreka, Steiner, Suess, Deecke, and Willmes, 1988; Roland and Friberg, 1985). These data provide strong support for the hypothesis that occipital lesions in rats made after acquisition would disrupt spatial orientation by destroying the site of storage of local views that are necessary for adapting behavior to the environment.

In conclusion, occipital lesions leave intact the perception of distances and simple objects which can be mediated by extrastriate brain structures such as the superior colliculus. The behavior required in response to the presentation of such a stimulus was "natural", such as jumping to a platform or orienting to a seed. Difficulties arise when more cognitively demanding and "artificial" visual tasks have to be performed such as the discrimination of complex stimuli or learning sets. In line with these conclusions, habituation of exploratory activity and reactions to spatial changes which rely on spontaneous behavior were unaffected by occipital lesions (Krimm, 1994). When tested in acquisition in the 3-table task or in the Morris swimming pool, deficits can be observed but they are not dramatic. Occipital-injured rats improve their performance level but they do not reach that of controls. When the lesion is made after acquisition, deficits in the radial arm maze and in the Morris swimming pool are much more profound, and persistent, showing no improvement of performance over time.

Impairment in retention may be accounted for if one assumes that occipital lesions made after training destroy the stored visual scenes necessary for other specialized brain structures (e.g., the hippocampus) to perform spatial tasks. In contrast, peripheral blindness would still allow animals the possibility of orienting on the basis of exclusively represented visual images. These could be used to match current non-visual experience (olfactory, auditory, and movement-related information) with a represented visual frame of reference. In rats having learned a spatial task while already blind, performance is left undisrupted by visual cortical lesions since the strategies which have been used during learning, before the lesion, have not involved any visual structure.

It could be argued that efficient navigation could be achieved on the basis of non-visual stored information. However, in most of the experimental situations that have been used, visual landmarks are the sole elements which allow the animals to deduce the location of a goal. In addition, even in rich multisensory situations, vision plays a prominent role in spatial knowledge and memory. This sensory modality is endowed with many advantages (amount, availability, simultaneity, and precision of information, for instance) over the others. Even though it is not the unique sensory channel whereby spatial knowledge is acquired, the value of distal and precisely located visual landmarks in the construction of representations is indisputable. No other sensory channel could provide equivalent information (Thinus-Blanc and Gaunet, 1996). Thus, during acquisition of a spatial task, vision may play a calibrating role in the organization of information provided by other sensory channels. Non-visual information would be processed and integrated within a salient and accessible framework provided by vision. I conceive the visual world as supplying a blueprint into which non-visual information is incorporated on the basis of the detection of concurrence of visual and non-visual information emanating from the environment. Subcortical visual structures such as the superior colliculus are equipped to perform this integrative multimodal function together with, or in the absence of, the occipital cortical areas. However, extrastriate visual processing is coarse and less precise than that performed by the occipital cortical regions. That may explain the relatively mild deficits of occipital rats during acquisition of a spatial task.

9. Concluding Remarks

In conclusion, damage to rats' prefrontal cortex results in a WM deficit which may take a variety of forms according to the nature of the task. As underlined by Granon and Poucet (1995) and Granon et al. (1994), the apparently incoherent results which are sometimes observed following prefrontal lesions "... can be explained if one considers the relative difficulty of each task: within each study, the working memory deficit is exacerbated in the task that is the most difficult for the rat to acquire ... This point merely suggests that the frontal cortical memory system does not function separately from other processing systems, such as the attentional processes required for acquiring complex associations." (Granon and Poucet, 1995, p. 482.) Concerning spatial tasks, the prefrontal area is conceived as a short term WM buffer maintaining the availability of spatial information necessary to plan an oriented trajectory. It is involved in the execution of spatially oriented behaviors but its function is much more temporal than spatial.

Visual cortical areas appear to contribute to the spatial organization of non-visual incoming input on the basis of a visual canvas that facilitates this processing. These would also be conjointly involved in both the perception and recall of visual scenes. It is speculated that this two-fold function could account for the deficits observed in the acquisition and retention, respectively, of spatial learning tasks in animals with occipital cortical lesions.

An important remaining question is how best to portray the interactions among the neocortical areas that have been examined and how they may cooperate with the hippocampus to ensure the chain of processes that results in complex and efficient spatial orientation.

CHAPTER 9

NEUROBIOLOGICAL MODELS OF SPATIAL PROCESSING

> *"Apparent differences between theories may not reflect a fundamental incompatibility at a functional level ... some apparent contrasts may reflect differences of focus and emphasis." (McClelland et al., 1995)*

One of the aims of this chapter is to present models which deal with the implementation by the brain of spatial processing and memory. Some of these models exclusively concern the hippocampus, either because they account only for one specific type of encoding (setting up of place-cell firing fields: Sharp; distance encoding: Muller and coworkers), or because the basic assumption is that the hippocampus can perform all of the stages in the process (O'Keefe). Other conceptions rely on the cooperation of the hippocampus with neocortical structures, such as the parietal cortex (Poucet; McNaughton and coworkers). Finally, four general models of memory are presented (Teyler and DiScenna; Miller; Damasio; McClelland and coworkers). Although these models are not specifically aimed at accounting for spatial processing, these are of interest in the present context: in Chapters 7 and 8, I have presented several data which strongly suggest that the associative parietal and prefrontal cortices, and to a lesser extent the occipital area, do subserve complementary spatial functions. Consequently, it is of importance to have some insight, or, at least, hypotheses about how their possible interactions and cooperation with the hippocampal system may be implemented within the brain. In this respect, theories concerning the neural substrates and evolution over time of memories in general provide interesting perspectives which can be applied to spatial memories in particular. Indeed, the memory components of spatial processing can be reasonably thought to be controlled by the same brain mechanisms as those involved in the memory components of any other type of processing (of comparable cognitive level).

The present review is far from exhaustive, since an exhaustive review would require a separate volume. The selection which has been made here was motivated by the concern to incorporate several levels and facets of spatial processing, from the setting up of hippocampal place-cell properties to vector computations that enable navigation in large scale environments. The models of general memory which have been chosen share common basic features and support each other, all the more so because the respective supporting arguments are drawn from rather disparate fields of research.

The idiosyncratic choice of models selected for presentation here should not obscure the contribution of other prominent conceptual frameworks that are not elaborated in the present volume to the understanding of memory in general, and spatial memory in particular (see, for example, Eichenbaum et al., 1994; Kesner, 1980, 1984, 1986; Rudy and Sutherland, 1995; Sutherland and Rudy, 1989; Worden, 1992).

Only the general outline and basic principles of each model are presented here. Detailed accounts of behavioral and physiological arguments which support them can be found in the articles and books cited below.

1. The Setting Up of Location-Specific Firing: Sharp's Model

The basic framework of Sharp's (1991) model to account for the establishment of place-cell firing fields is a neural network with the characteristics of a competitive-learning, pattern-classification device (e.g. Rumelhart and Zipser, 1986; Von der Marlsburg, 1973). It comprises three layers of neuron-like elements. Each element of one layer can be connected to each element of the next lower layer by Hebb-like synapses. The first layer is an input layer of neocortical cells corresponding, in the model, to polysensory association cortex. The second layer is the entorhinal cortex which receives massive inputs from polysensory association areas such as the temporal cortex (e.g., Van Hoesen and Pandya, 1975, in the monkey). The entorhinal cortex is also the main source of neocortical input to the hippocampus which corresponds to the third layer in the model. These last two layers are divided into "winner-takes-all" clusters. This design implies the firing of only one cell in each cluster, that cell which receives the largest input from the first layer and which, consequently, has the strongest connections with currently active cells. The firing established in the input layers projects to the second layer (entorhinal cortex), which itself projects to the third layer (hippocampus) according to the above principle.

Sharp's model rests on the following principles: while the animal is moving around within a given environment, various local views are available from each of its locations. Local views available from contiguous locations or from different directions within the *same* location are very similar since they share many elements in common. Consequently, they will be placed into the same class. A crucial factor which determines the grouping of two patterns or local views is the number of intermediate local views between them. Sharp writes:

"*Thus, if the two views available when the animal is facing two opposite directions in a given location are experienced in addition to other views available from intermediate directions within that location, the likelihood that those two opposite views will be placed into the same category is increased ... Alternatively, if no views of intermediate directions are available, the views from the local region for each of the two directions may each form their own "cluster" in the input space, and so they would be less likely to be placed in the same category."* (Sharp, 1991, p. 108.)

The function of the pattern-classification device is to classify and store local views on the basis of their similarities. According to its competitive-learning features, the model results in one given cell (a place-cell) firing in one precise region of space (a place-field) where similar local views are available. Even if local views perceived facing in opposite directions, but from the same location, are not similar, the continuity ensured by a succession of overlapping intermediate local views ensures their classification into the same category. Accordingly, a discontinuity (Fig. 9.1) is likely to result in the appearance of two distinct firing fields.

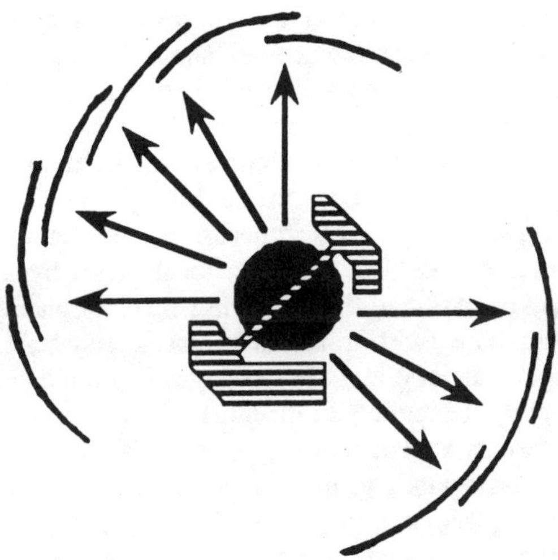

Fig. 9.1. Example of a case where a complete rotation on the spot does not provide a continuous perception of the overlapping local views. The striped blocks represent visual occluders. This configuration may encourage the setting up of two distinct firing fields.

According to Hebb-like synaptic principles, all connections which were activated when the place-cell fired are strengthened. Therefore, they are more likely to fire during subsequent presentations of one of the local views belonging to the same class, i.e., that are available from one location, regardless of the animal's position at that location.

Trajectories taken by fictional rats, similar to those taken by rats in the course of real exploration, are simulated within a cylinder similar to that used by Muller, Kubie and Ranck's group in actual recordings (see Chapter 6). Samples are taken at the theta rhythm frequency at which pyramidal cells fire. Such a simulation, according to the principles enunciated above, leads to firing-rate maps which are strikingly similar to those found in actual recording experiments in the same apparatus (Muller et al., 1987). In addition, the properties of the firing-fields share the same properties as in the actual experiments. For instance, the fields are stable over time (Best and Thompson, 1989; Muller et al., 1987), persist if some of the environmental cues are removed (O'Keefe and Conway, 1978; O'Keefe and Speakman, 1987) and have directional correlates if the fictional rat's displacements are constrained along determined paths (as in the radial maze) (McNaughton et al., 1983). Furthermore, their size is increased correspondingly with enlargements of the test apparatus (Muller and Kubie, 1987).

Finally, Sharp's model leads to interesting predictions. Sharp assumes that the likelihood that two locations (or views within locations) with some sensory overlap are represented by the same place-cell should be highly dependent on the amount of exposure to the intermediate scenes between them. The experimental situation designed by Save et al. (1996b, see Chapter 2) brings experimental support to this hypothesis at the psychological and behavioral levels. Save et al.'s data suggest that the critical factor for setting up spatial relationships is not the opportunity to travel along direct trajectories between two subspaces but rather the amount and *continuity* of available local views from locations within each subspace. It may be expected that the firing fields of place cells recorded in each of these four situations (with transparent-closed, transparent-open, opaque-closed, or opaque-open barriers) and after changes from one situation to another should display variability depending on the situation currently experienced.

This model represents a plausible explanation for the generation of place-cell firing-fields and corresponds to the setting up of a truly abstract processing of place as reflected by place-cell properties. A strong emphasis is placed on the necessity for continuity among overlapping local views, i.e., on the importance of exploratory activity for spatial knowledge as has been demonstrated at the behavioral level.[a]

---

[a] An important element which is lacking in this model is the integration of proprioceptive and vestibular information generated by the subject's rotation on the spot, together with the corresponding flow of visual scenes. Data about place-cell reactivity to movement-related information (see Chapter 6) have accumulated since 1991 and could be incorporated into an updated version of Sharp's model.

2. Encoding of Connectivity between Firing-Fields; The Hippocampus as a Cognitive Graph: Muller and Coworkers' Model

Muller, Kubie, and Saypoff (1991a) and Muller et al. (1991b) propose a theory which accounts for the way in which the CA3 place-cell population could come to encode the connectivity of the environment in a "holistic" manner. The theory is founded upon several place-cell properties. The most important property is that CA3 pyramidal cells make excitatory monocontacts with each other (Miles and Wong, 1986) and the strengths of these synapses increase along the lines of a Hebbian mechanism, when the pre- and post-synaptic cells fire in close temporal association (Miles and Wong, 1987). Another property is that distinct sub-sets of place-cells are activated for distinct environments. When the animal is placed for the first time in unfamiliar surroundings, a new active sub-set is generated, very soon after the rat's placement in the novel spatial situation (Hill, 1978; see Chapter 6).

The basic idea of the model is that synaptic strength between two place-cells (which belong to the same-subset activated in a given environment) is a function of the distance between their firing fields while the animal explores the environment. Fig. 9.2a represents a circular recording chamber where three cells are recorded. The firing fields of A and B overlap, whereas that of C is distant from both A and B. Cell A is monosynaptically connected to both B and C. While the rat is moving around in the chamber, it is expected that cells A and B will fire within a brief interval because the places where they fire are close to each other. In contrast, the delay between the firing of cell A and C will be far longer, all the more so if rats do not move quickly. Accordingly, the strength of the synapse from A on to C will be weak.

This basic hypothesis has been tested by computer simulation (see Muller et al., 1991a, for details of the computations), on the basis of coordinates of the successive positions of an actual rat chasing food pellets within a circular chamber. Fig. 9.2b shows the relationship between synaptic strength and distance between the firing field centers. Strength is all the more important when distance between fields is small. Since each synapse was considered as an independent storage unit, varying the number of cells (between 100 and 3600) did not modify the profile of the relationship between synaptic strength and field distance.

On this basis, the authors propose that this type of spatial representation is a "directed graph" because the flow of information is unidirectional along the axons and across synaptic relays. Cell bodies and dendrites would be the nodes of the graph, and the axons would represent edges. The structure of this encoding in the brain might be functionally isomorphic to that of the structure of the environment, and this isomorphism would allow for the use of distance representations.

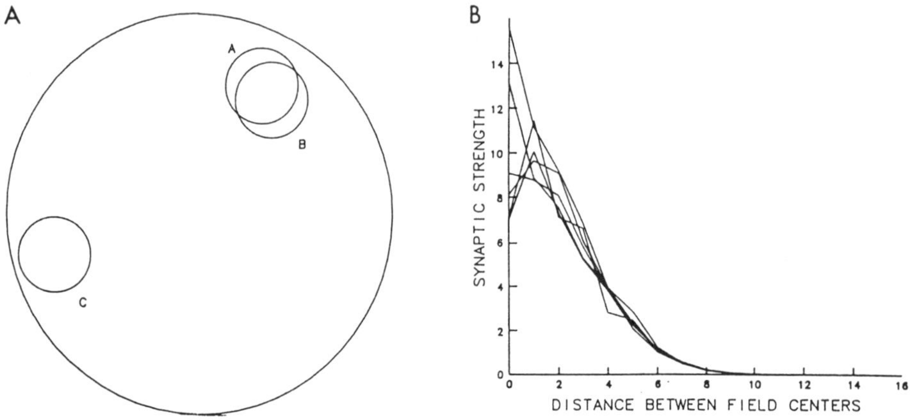

Fig. 9.2. A: Idealization of the spatial arrangement of the firing fields of three CA3 place cells in the cylinder. B: Synaptic strength as a function of distance between field centers with networks containing different numbers of cells. (From Muller et al., 1991a.)

One of the advantages of the model is that since a small proportion of cells is active in a given environment, it is possible that representations remain independent and that there is no interference among the representations of various situations. Moreover, it opens promising perspectives for studying spatial computation, in particular for generating new solutions to spatial problems such as making shortcuts and detours on the basis of encoded connectivity. It should be mentioned, however, that this model accounts only for short distance encoding since there is a distance-threshold where synaptic strength between two cells becomes equal to zero. Consequently, the model does not account for distance encoding beyond that critical threshold.

3. The Centroid Model for the Hippocampal Cognitive Map: O'Keefe

The model recently proposed by O'Keefe (1988, 1990, 1991a and b) is not at odds with the scenario developed in the initial hypothesis (O'Keefe and Nadel, 1978; see Chapter 1; Nadel, 1991) but, rather, represents an extended and more precise account, in computational terms, of how environmental information could be processed by the hippocampal "mapping" system so as to be used for navigation.

The first step in the processing concerns environmental cues. These are individually located within an egocentric space each having a distance and angle. According to O'Keefe, the mapping system, in this first stage, computes the geometric center or *centroid* of the set of cues in the allocentric space. The average of several cue vectors provides a vector C that identifies an unoccupied place in the environment which is the centroid of the cue array (Fig. 9.3a). When the animal makes a rotation on the spot, the length of the centroid vector does not change whereas both its length and angle vary with translation movements. These vector variations can be taken as reflecting an updating of the animal's successive positions by reference with the stable centroid location.

The second computation performed by the mapping system is that of the *slope* of the distribution of the set of cues. The line that can be drawn between each pair of cues has a slope that is computed on the basis of the x and y coordinates of each cue (Fig. 9.3c). The average of the slopes for all of the cue pairs results in an overall slope or overall direction, that is the second main element required for the mapping system to be functional.

Unlike the centroid vector which does not vary with rotations on the spot but varies with translational movements, the animal's orientation with respect to the slope axis varies with rotational movements but not with translations, i.e., it does not change in different locations provided that the animal's head points in the same direction.

According to O'Keefe, the mapping system has

"... *two independent measures derived from the environment, the direction, which is independent from the animal's location in the environment, and the location, which is independent of the direction. Together, they provide an allocentric reference framework, which fulfills all of the properties of a cognitive map.*" (O'Keefe, 1991, p. 232.)

The theory assumes that the computation of both the centroid and slope are performed by the hippocampus. First, the CA3 place-cells make an estimate of the centroid on the basis of the subset of cues that is available to each of them. O'Keefe proposes that vector encoding relies on the sinusoidal properties of the theta system by a phasor mechanism used to encode the angle and distance to a landmark. The notion of phasor refers to the fact that any vector in polar coordinates can be represented as a sine wave with a defined amplitude and frequency, the former representing the length of the vector (or distance to the cue) whereas the phase of the sine wave relative to a reference represents the angle in egocentric space (Fig. 9.3b).

"*The theta system is viewed as a compound waveform (Vanderwolf and Leung, 1983), the function of one component being to provide the standard sinusoidal reference wave against which phase angles are measured. Vector addition in such a system is performed by linear summation of the sinusoids.*" (O'Keefe, 1991, p. 232.)

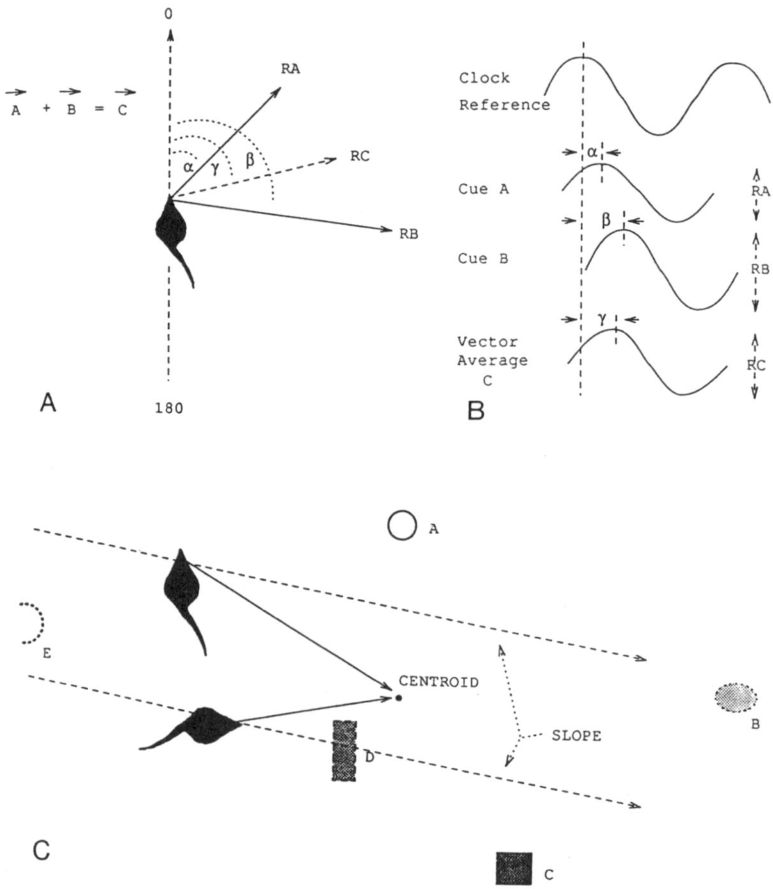

Fig. 9.3. Calculation of the centroid and slope of an environment on the basis of egocentric cue information. A: The representation of two cues in an egocentric polar framework centered on the head. Each cue is represented by a vector (A and B), whose lengths RA and RB are related to the distance of the cues and whose angles (alpha and beta) correspond to the angles in the head framework. Averaging of two or more cue vectors yield a vector (C) that points to the geometrical center (centroid) of the cue array. B: One way in which the vector calculation might be performed in the hippocampal system. In phasor notation, each cue vector can be represented by a sinusoid in which the amplitude of the wave is related to the length of the vector RA, and the phase shift of the wave relative to a clock wave is related to the vector angle (alpha). The linear average of the sinusoids gives the vector average of the vectors. C: The centroid and the slope of an environment with five cues, A-E. The length of the centroid vector is invariant with rotations by the animal in one location but varies with translations. The angle between the slope and the rat's egocentric axis is invariant with translations but changes with rotations. Together, the centroid and the slope provide the basis of an allocentric coordinate framework. (Reproduced from O'Keefe, 1991a.)

The computation of the slope is based on head-direction cells in the dorsal presubiculum (see Chapter 6). Movement-related inputs are used to update the animal's current heading by reference to the overall direction, and translation and rotation movements are also represented as vectors carried by a second component of the theta system (for details, see Burgess, O'Keefe, and Recce, 1993; O'Keefe, 1990, 1991a abd b; O'Keefe and Recce, 1993).

Although O'Keefe and coworkers' hypothesis about sine wave and vector encoding has raised critical comments and alternative accounts (Touretsky, Redish, and Wan, 1993), the fact remains that the model leads to interesting predictions, some of which have been confirmed. For instance, it predicts at the behavioral level, that any spatial modification that affects the centroid and the slope of the current representation of an environment will induce re-exploration (occurring in order to update the initial representation by incorporating the change). Conversely, symmetrical changes in object locations leading to uniform enlargement or compression of a cue configuration should not elicit re-exploration of the situation. That is exactly what occurred in the dishabituation experiments with objects in the field (Poucet et al., 1986; Thinus-Blanc et al., 1987) presented in Chapter 2. Enlarging or decreasing the size of the square defined by the objects did not elicit any exploratory reactions in hamsters. In contrast, modifications of the shape – which modified the location of the centroid and the slope – in keeping with O'Keefe's formulation, led to a renewal of exploration. Indeed, this prediction could also be tested at the electrophysiological level by examining modifications of place-cell firing fields following a similar manipulation of cue-objects inside an open field.

4. Implementation at Brain Level of McNaughton's Conditional Association Model

Associative representations between specific movement and local views, proposed by McNaughton (1989a; see also McNaughton and Nadel, 1990; McNaughton et al., 1989b) are represented in the brain in the form of a correlation matrix. This, in turn is implemented in neural circuitry as a modified version of the Hebb-Marr model. This formulation accounts not only for recall, but also for pattern completion, i.e., for the fact that the presentation of one of the elements stored in memory is enough to activate all the other elements which have been associated with it (see McNaughton and Morris, 1987, for a detailed account of this aspect in relation to Long Term Potentiation).

Several sets of neurons have to cooperate in the working hypothesis represented in Fig. 9.4. The combination of the output of a heteroassociative matrix corresponding to "place" (A) representation with the output (L) of the "movement-representation" system results in the unique compound representation (AL) of a left movement executed from A. By projecting this place-movement association back to the matrix via modifiable synapses, this becomes associated with its immediate result, i.e., the representation of location B. This location would subsequently be recalled each time the compound representation AL is presented (pattern completion).

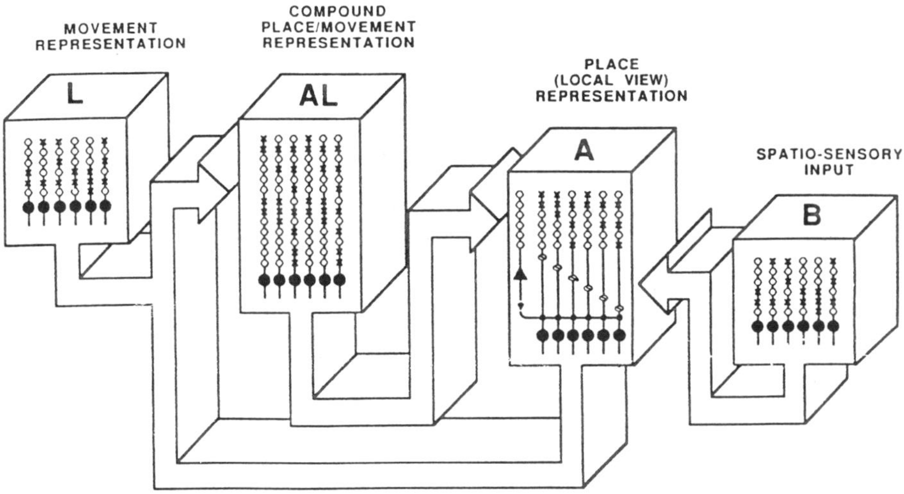

Fig. 9.4. A simple model for implementing the transition matrix idea for conditional associations of local view representation with movements. (Reproduced from McNaughton, 1989a.)

Such a mechanism only requires that the starting location has been presented, in order for chains of representations of locations to be acquired on the basis of movements which link them. One of the arguments in support of the theory is that place-cells continue to fire in darkness provided that the rat "knew" its starting position (e.g., O'Keefe and Speakman, 1987).

Various categories of neuron ensure this process. Briefly, hippocampal pyramidal cells would be involved in place (local view) representations. McNaughton and coworkers do not refer to place-cells which fire regardless of the animal's position but to "local view" cells. The firing of these cells, which have been identified in the radial maze, depends not only on the animal's location but also on the direction in which it is facing (McNaughton et al., 1983).

Many cells found in the posterior parietal cortex of the rat appear to fire according to the animal's movements (e.g. turning right, turning left, or moving forward) and also require the presence of some visual landmarks or intra-maze features (Chen and McNaughton, 1988; McNaughton, Green, and Mizumori, 1986a). Such units are taken by the authors as reflecting the representation of specific configurations of movements and local views. Place-motion configurations would become associated with the result of the movement, i.e., with the next available location (local view). Finally, head-direction cells (which have been found in various parts of the brain – the subiculum, posterior cortex, thalamus, and striatum; see Chapter 6) would maintain an "inertial direction sense" by integrating information (from the parietal area) about angular movements. Such a directional sense would allow for the computation of new trajectories, such as shortcuts, to be made on the basis of initially established movement-location associations.

Thus, parietal neurons are assigned in this conceptual framework a specific and important function in spatial encoding, in cooperation with the hippocampal formation. McNaughton et al. (1989b) suggest that this view may contribute to a better understanding of how the parietal and inferotemporal pathways, (the "where" and "what" systems respectively, demonstrated in primates by Ungerleider and Mishkin, 1982) might be reintegrated to ensure visuospatial processing (see also section 5).

The model proposed by McNaughton and coworkers has been the subject of criticism (e.g., Muller et al., 1991b; O'Keefe, 1991b) though its empirical bases have not been questioned. In my view, the aim of McNaughton and coworkers is not to account for cognitive mapping but for "the representation of action in space" (McNaughton and Nadel, 1990). In this respect, the model proposes one *possible* representation of action in one *specific* type of space, namely mazes with alleys (where directional firing of place-cells has been observed). In no way is it generalizable to navigation in open environments. In addition, though the authors use the term "location", what is being referred to are local views, i.e., visual scenes, the identification of which depends on a given direction of approach. Consequently, the most prominent and interesting properties of place-cells, namely their omnidirectional firing, is not incorporated into the model.

More recently, McNaughton, Chen, and Markus (1991) and McNaughton, Knierim, and Wilson (1993a) have proposed a more elaborated model which basically relies, at the psychological level, on Collett et al.'s vector computation (see chapter 5). The vector C (computed trajectory to the goal; Fig. 5.1) results from calculating the difference P (perceived vector to landmark A) less S (stored vector). This model is more adapted than the previous one to navigation in open environments though based on very similar empirical arguments.

As in the Conditional Association model, a directional reference system is provided by head-direction cells. The integration of information from the vestibular system by head-direction cells is formalized by McNaughton et al. (1991). These authors propose a computational model that accounts for the new bearing that follows a rotation of a fixed number of degrees, by reference to an initial bearing. In order to avoid cumulative errors arising from the integration of successive movement-related inputs, information from the environment must also be incorporated into the system. McNaughton et al. (1991, 1993a) suggest that correction may be achieved by associations between head-direction cells and local view cells.

Unlike the Conditional Association model, McNaughton et al.'s more recent account of the computation of trajectories relies on vector computation. Consequently, there must be, in the brain, cells which ensure vector encoding. The authors assume that both hippocampus and parietal cortex place-cells may play this role:

"*The proposition that a cell is location-specific, irrespective of egocentric orientation, is equivalent at least until tested to the proposition that it encodes the location and bearing of at least one fixed landmark.*" (McNaughton et al., 1993a, p. 591.)

According to the authors, each place cell represents a vector, the bearing of which is set by head-direction cells, distance being derived from various types of visual information (stereopsis, motion parallax, and so on).

Another requirement is that there is a brain mechanism which permits vector subtraction. McNaughton et al. (1993a) propose a solution based on "associative mapping principles". Each possible P × S combination is mapped into one value of C and all of these combinations have to be encoded by specific units. The computation of C is performed by the associative mapping network on the basis of both P and S vectors.

On the basis of studies that showed that the primate posterior parietal cortex encodes interactions between retinal location and eye position so as to transform retinal coordinates into head coordinates (Andersen, Essick, and Siegel, 1985; Zipser and Andersen, 1988), McNaughton et al. (1993a) conclude that, provided that coordinate transformation is written in vector notation ... "it becomes clear that vector subtraction is not something unknown to parietal cortex".

Some aspects of this model are highly speculative, such as the generalization to other species of complex coordinate transformations performed by the monkey parietal cortex. This is the weakest point of the model. Though it accounts for some existing data and makes plausible and interesting predictions for new experiments (see McNaughton et al., 1993a), the hypothesis about how the brain might perform vector subtractions remains unclear. In contrast, the existence of a directional sense serving as a reference for integrating changes of bearings on the basis of vestibular input has recently received experimental support (Goodridge and Taube, 1995; McNaughton et al., 1993b). The identification of a directional system, acting as a reference for the estimate of orientation changes and integrating these changes, is the element that was missing in the behavioral version of the Vector model proposed by Collett et al. (1986). More generally, the conceptual framework delineated by McNaughton involves not only the hippocampal system but also the parietal area. This parieto-hippocampal, and, in a wider context, cortico-hippocampal interplay is developed further in the model proposed by McClelland et al. (1995; see section 9) which concerns memory in general.

5. The Neural Substrate of Poucet's Model of Spatial Processing

The implementation of Poucet's model by the brain encompasses three structures: the hippocampal formation, and the parietal and prefrontal cortices. A special emphasis is placed on the electrophysiological properties of neurons of the hippocampus and associated structures.

At a general level, the neural circuitry for spatial information processing is based on that proposed by Mishkin, Ungerleider, and Mako (1983) and Ungerleider and Mishkin (1982) for visual outflow in primates. This general circuit had been discussed previously by McNaughton et al. (1989b). Basically, the circuitry consists of two cortical pathways originating from the occipital cortex. The occipitoparietal pathway (OP) carries information about the spatial location of visual stimuli whereas the occipitotemporal pathway (OT) conveys elements necessary for object and shape recognition (Fig. 9.5). Poucet assumes that these two pathways converge onto the entorhinal cortex which, thus, receives information issued from each type of specific processing.[b]

The various stages of spatial information processing put forward by Poucet are inscribed within this general framework. The first stage is that of the setting up of place representations (location dependent reference frameworks) resulting from the integration of the various local views available from one single place during rotational movements. This function would be ensured by hippocampal place-cells along the lines of the computational model proposed by Sharp (1991; see section 1). The second stage consists of establishing connections between place representations. The hippocampus, in particular place cells from the CA3 field, would ensure this encoding along the lines of Muller et al.'s (1991a) model (see section 2). Locations that are the object of place representations (i.e., where a place cell is active) are linked by topological relationships, such as proximity and connectivity on the basis of the strengthening of synaptic weights between place cells.

One of the original features of the model is that instead of using one unique directional reference, as long as navigation takes place within local charts (where landmarks are available from both starts and goals), animals compute

---

[b] Although this circuitry was initially described in primates, some of its connections have been demonstrated in the rat (see Poucet, 1993, for details).

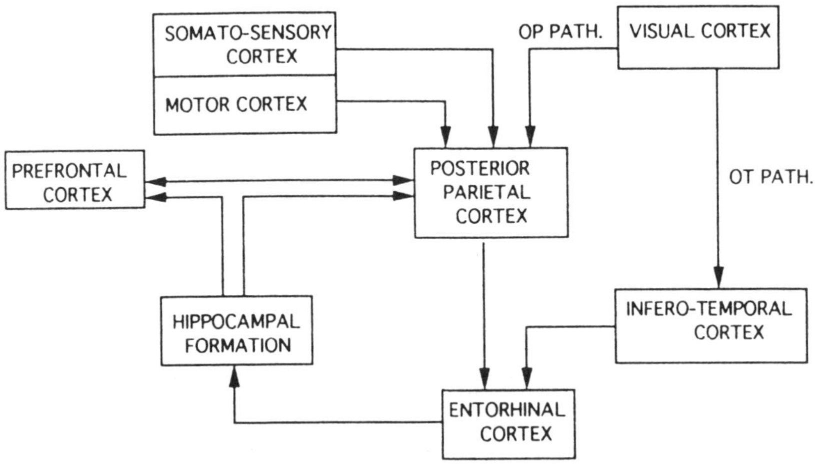

Fig. 9.5. A schema of the hypothetical circuitry involved in spatial processing. (Reproduced from Poucet, 1993.)

several location-dependent reference directions which allow them to determine directions from any particular location. The population of postsubicular head direction cells would provide a full 360 range of available reference directions:

"*The postsubiculum therefore would serve as a database system providing information useful for computing directions. If we assume that such directional information can be combined with the locational information provided by place-cells, then the combined system provides full information for further processing by other systems*" (Poucet, 1993, p. 178).

Among the other systems, the posterior parietal cortex would combine topological information from the hippocampal formation with movement-related information so as to "enhance the grain" of the rough hippocampal representation by providing it with metric properties.

Lastly, the prefrontal cortex would be involved in the computation and planning of complex trajectories because of its working memory function (see Chapter 7). Poucet proposes that this area may receive topological information from the hippocampus and/or finer grained metric information from the parietal cortex.

It should be emphasized that Poucet does not provide any account of the implementation by the brain of the ultimate stages of his model. For instance, the mechanism whereby each block of location-dependent information is combined into a unified location-independent representation, with one unique reference direction, is not considered. Any solution would be speculative at this stage.

In spite of this shortfall, Poucet's model has several advantages. Besides the general interest which arises from considering the respective contribution of the hippocampus and other neocortical areas, it accounts, among other things, for the sparing of spatial abilities following hippocampal damage occurring a relatively long time after training. The model assumes hippocampal involvement during the initial acquisition but, subsequently, information would be stored in both the hippocampus and parietal cortex. Because of this partial redundancy in the stored spatial information, and provided that the goal location, for instance, has not changed, the parietal cortex could ensure accurate navigation following hippocampal damage.

In addition, encoding topological relationships in terms of synaptic strength means that this strength would fall to zero when locations are distant (see section 2), accounting for the dissociation proposed by Poucet between long range and short range navigation. Finally, another interesting prediction is that hippocampal damage should affect tasks which rely on topological encoding whereas parietal lesions should interfere with metric encoding. The typical heading errors of parietal rats tested in the Morris swimming pool represent a preliminary confirmation of the prediction: these animals are able to roughly locate the platform but cannot precisely adjust their trajectory toward it.

6. The Hippocampal Memory Indexing Theory: Teyler and DiScenna's Model

The key idea of Teyler and DiScenna's theory (1985, 1986) is that "... the pattern, or index, of specific neocortical (and other) areas activated by an experiential event is represented, or indexed, in the hippocampus by means of the LTP" (1985, p. 377). According to this view, the hippocampus is not the site of storage of memories but rather holds "addresses" of neocortical regions that, when reactivated, recreate the representation of the initial experiential event.

The phenomenon of long term potentiation (LTP) can be defined as a long-lasting increase in synaptic efficiency following afferent fiber tetanization. In addition to the various studies showing LTP consequent upon artificial stimulation, other studies have shown that LTP also occurs spontaneously. For instance, Sharp, McNaughton, and Barnes (1985) have demonstrated a spontaneous synaptic enhancement in rats exposed to a spatially complex environment. Several studies of LTP support the idea that this can be considered a good candidate for the maintenance of memory (for reviews, see Cotman and Lynch, 1989; McNaughton, 1983). LTP could maintain information in the hippocampus about stimulation arising from other parts of the brain.

This information would be topographically organized within the hippocampus. Arguments supporting this assertion come from studies showing that the claustrum, which has a reciprocal connection with the entorhinal cortex, receives from the neocortex projections which reflect, to some extent, its topographic organization (Carmen, Cowan, and Powell, 1964; Olson and Graybiel, 1970, quoted by Teyler and DiScenna, 1985). It is also widely acknowledged that entorhinal projections to the hippocampus are topographically organized (Hjorth-Simonsen, 1972; Hjorth-Simonsen and Jeune, 1972; Steward, 1976, quoted by Teyler and DiScenna, 1985). However, the topographical nature of the projections from the claustrum to the entorhinal cortex has yet to be demonstrated.

Another explicit feature of the theory is that the associations of the initial experiential events are mediated by LTP occurring between cortical modules. The hippocampus holds only the index or spatio-temporal pattern of the cortical modules which have induced hippocampal LTP. The links between several experiential events are stored at cortical level whereas each individual event is represented in the hippocampus by means of the hippocampal index.

When only some of the indexed neocortical modules are activated, the hippocampal memory index will also be partially activated. However,

"*If the hippocampal activation exceeds some threshold level, the remainder of the hippocampal index will be activated and will, in turn, reactivate the entire neocortical array. By this means, a partial experiential event, given that it is sufficiently complete to exceed a threshold for hippocampal memorial recognition, can be recognized as a memorial event by its ability to recreate the pattern of neocortical modules activated by the original experiential event.*" (Teyler and DiScenna, 1985, p. 384.)

An additional feature of the model is that repetitions of an experiential event may strengthen the circuits between cortical modules. If so, there is redundancy between these circuits and the hippocampal index. This would explain the well-known fact that retrograde amnesia following hippocampal damage in animal and human subjects often spares ancient memories. The role of the hippocampus as an index would be transiently crucial, mainly during acquisition and during a more or less extended subsequent period.

The relevance of this general neurobiological model of memory for spatial processing is partly supported by experimental evidence. On the one hand, the disruptive effects of hippocampal lesions on spatial performance in acquisition and retention can be satisfactorily accounted for in terms of the Memory Indexing theory. The neocortical regions involved in spatial processing could not be "registered" into the missing hippocampal index when a lesion was made before training (acquisition). Accordingly, no representation could be constructed. In the case of post-training surgery (retention), addresses of the neocortical loci involved during learning would be destroyed. The spatial representation could not be reconstructed unless a sufficient delay between the end of learning and the lesion has allowed for strengthening of circuits between the neocortical areas involved in learning.

Thus, the theory provides a reasonable account of the memory side of spatial processing. In contrast, it does not consider how spatial features may be encoded by the hippocampus, and it does not even allude to place cells and to their properties. In this respect, it cannot be easily applied to the encoding and computational aspects of spatial processing.

Whether or not Teyler and DiScenna's model is relevant for spatial processing, it suffers from an important weakness: two fundamental pre-requisites for its plausibility, i.e., the mapping of the activity of neocortical areas upon the hippocampus, and the neural substrate for the hippocampal index, have yet to be clearly demonstrated. Indeed, the prominent characteristic of the theory – which made it very influential – is that the authors have outlined a precise, though speculative, scenario of cortico-hippocampal interplay in which the hippocampus plays a key role. This idea is found in other models which have been more recently advanced (see next sections).

7. The Cortico-Hippocampal Interplay Model: Miller's Model

Miller himself summarizes his model in the following terms:

"*... Hebbian processes of synaptic strengthening select patterns of loops passing from hippocampus to cortex and back again to the hippocampus. The total conduction delay time round each one of these patterns of loops is envisaged to correspond to the theta period. Such resonance between hippocampus and cortex is envisaged to have an important functional role in registration and retrieval of information in the cortex. Each pattern of resonant loops will raise the activation of specific collections of cells which are widely dispersed across the cortical mantle. Such collections of cells have insufficient mutual interconnections within the cortex to form a cell assembly by the usual Hebbian mechanism ... Nevertheless, they do represent definite information groupings. In fact, it is envisaged that they represent large-scale aspects of the environment as a whole, those we designate as contexts.*" (Miller, 1991, p. 159.)

First, Miller assumes that the "local cell assemblies" which are found in neocortex do not solve the problem of ambiguity. An object presented without taking into account its context is likely to be erroneously identified. The neocortex is not equipped for detecting ambiguities. If such were the case, the number of necessary connections would become biologically implausible. "Global cell assemblies", holding information about an object within a specific context, are unlikely to be formed within the neocortex (see Miller, 1991, Chapter 2, for arguments supporting this assertion). Consequently, a global cell assembly requires the coordination of the activity of local cell assemblies scattered throughout the cortex. In addition to the Hebbian principle ensuring the strengthening of synapses within local cell assemblies, organizing mechanisms are required for the recall of ambiguous or complex situations.

These organizing mechanisms are assumed by Miller to be found within the hippocampus on the basis of arguments related to the presence in this structure of rhythmical slow activity (RSA) or theta rhythm. The occurrence and frequency of RSA is highly dependent on the species under study, the state of the subject during recording (freely-moving, urethanized or curarized) and the task performed. This makes it very difficult to arrive at an unitary conclusion from the diversity of data. For instance, RSA in the rat has been shown to be correlated with any movement involving the whole body (Vanderwolf, 1969), including swimming (Whishaw and Vanderwolf, 1971), or with small movements from a fixed base, such as head movements (Whishaw and Vanderwolf, 1973). Other studies have corroborated and extended the correlations obtained by Vanderwolf's group (cf. details in Miller, 1991, Chapter 4). In some studies, RSA has also been found to be associated with activities related to the gathering of information from the environment, for instance during the animal's confrontation with a novel situation (Bennett, 1970).

After having thoroughly discussed some of the respective properties of the hippocampus and of local cell assemblies in the neocortex, Miller defines the nature of their interplay on the basis of patterns of self-organized phase-locked loops entrained by hippocampal theta. Each pattern encodes a particular environment. Therefore, specific context representations are available. The ambiguities inherent in the information structure of the environment at the level of the cortical network would be resolvable by this means.

"*The hippocampus thus acts in some ways as an 'index' to memories, the cortex being, as it were, the 'book' to which this index refers.*" (Miller, 1991, p. 159.)

Although the process of retrieval is not explicitly developed by Miller, it can be easily inferred from the theory as it stands. The activation of a local cell assembly (belonging to a pattern of loops involving several local cell assemblies and corresponding to a given environment) would allow the retrieval of the whole environment (i.e., the co-activation of these local cell assemblies) via the incorporation of the hippocampal "index".

Miller's hypothesis has several strong requirements at brain level. First, in addition to the demonstrated reciprocal connections between the hippocampus and neocortex, the time required to complete the total loop between them must fit the frequency of the theta rhythm. Indeed, there exists a wide range of delay lines between the hippocampus and the neocortex because of different axonal lengths and diameters. However, according to Miller, it is sufficient that only a small proportion of recursive loops match the theta period. What is important is that those loops are selected.

This remark leads to the second pre-requisite: What is the mechanism allowing for resonant circuits to be selected among the wide range of possible resonant circuits between the hippocampus and neocortex? Synaptic strengthening has been shown to occur according to Hebbian or quasi-Hebbian rules in both neocortex (Fregnac, Shultz, Thorpe, and Bienenstock, 1988, for instance) and hippocampus (Goddard, 1980; Wigstrom and Gustaffson, 1985). On this basis, Miller proposes the following scenario:

"*Imagine that the hippocampal oscillator comes into action at the same time as a group of isocortical cells receives tonic activation from the environment: the cortical neurons will be activated tonically, and will also tend to receive rhythmic signals via connections from the hippocampus. This is a condition in which the Hebbian principle can operate. When it does so, the converging synapses relaying the rhythmic activity will be progressively strengthened. As a result of this, the group of cortical cells would show not only tonically elevated activity, driven by environmental stimuli, but superimposed on this, a rhythmicity phase-locked to the hippocampal rhythm. Pathways returning from these rhythmically active cortical cells to the hippocampal oscillators would therefore come to transmit similar rhythmic signals.*" (Miller, 1991, p. 163.)

These reciprocal connections have a delay which approximately matches the period of the hippocampal theta rhythm. Therefore, there is a coincidence of pre- and post-synaptic activity at a given moment of the theta cycle for the hippocampal synapses of these connections. This meets the criterion for Hebbian synaptic enhancement. Resonance can then occur on the basis of synaptic strengtheening. The same mechanism also applies to the intermediate synapses involved in the resonant loops.

One of the weaknesses of the model is that theta activity is not widely detectable in many regions of cortex connected with the hippocampal formation. More importantly, there appears to be no phase correlations between hippocampal and cortical RSA. Only one study (Mitchell and Ranck, 1980) demonstrates that theta activity in the entorhinal cortex is approximately in phase with that of CA1 in the hippocampus, in the freely moving rat. But the possibility that entorhinal theta activity was volume conducted from the hippocampus rather than generated within the entorhinal cortex cannot be discarded (cf. Miller, 1991, Chapters 6 and 7). Indeed, as emphasized by Miller, no studies have been conducted within a conceptual framework within which the precise hypotheses generated by his model can be tested.

Like Teyler and DiScenna's theory, Miller's model attempts to account for the sparing of ancient memories following hippocampal damage. Miller envisages that once cortico-hippocampal circuits have been established, many neocortical networks may become entrained on the rhythm imposed by these loops. Consequently, the setting up of cortico-hippocampal circuits plays a "priming" role in establishing the spatial and temporal structure of neocortical activity. Once this is established, the hippocampus may not be necessary for retrieval.

Though Miller's model is more extensive than that proposed by Teyler and DiScenna, it is based on the same distinction between, on the one hand, many somewhat independent elements of a given memory (stored in neocortex), and, on the other hand, one single organizing structure, the hippocampus, co-activating these neocortical areas. The main difference lies in its incorporation of more data than Teyler and DiScenna's memory indexing theory and it accounts for the cortico-hippocampal interplay with reference to different mechanisms.

Concerning the relevance of Miller's model for spatial cognition, it shares some of the advantages and drawbacks of the Memory Index theory. It does not concern spatial encoding but rather the memory aspects of spatial processing. Like Teyler and DiScenna's model, it accounts for the differentiated effects of hippocampal lesions in the acquisition and retention of spatial tasks. An additional interesting feature relating to the putative function of the hippocampus for co-activating "local cell assemblies" so as to form "global cell assemblies" is that it renders unambiguous complex situations while at the same time it reduces memory load. One local cell assembly corresponding, for instance, to the visual image of an object – or to a local view – can be part of several situations or contexts depending on the other local cell assemblies co-activated by the hippocampus. This notion is germane to O'Keefe and Nadel's assertion that the information held in cognitive maps is "context-specific" since the representation of a place on a map is encoded in terms of its spatial relationships with other elements; no confusion is possible between similar elements belonging to different situations.

8. The Time-Locked Multiregional Retroactivation Theory: Damasio's Model

One of the original features of Damasio's theory (1989a and b, 1990) is that various levels of integration or "convergence zones" are taken into account. Consequently, the hippocampus, although assigned an important role, is not the keystone of the theoretical edifice. The other distinctive feature is that the model relies on arguments which are totally different from those supporting the two preceding models. They are drawn from clinical observations of brain damaged humans and from lesion studies in primates.

Damasio's theory can be summarized as follows. Memories are not stored in extenso. Rather, the elements of the memory are separately stored in the sensory and motor brain areas which have participated in the processing of the experienced situations. "Convergence zones" contain the "combinatorial or binding code" allowing the simultaneous retroactivation of the elements forming the memory. They are created when they are co-activated by inputs coming from various neocortical and subcortical regions and they hold the pattern of temporal and spatial coincidences of the elements of the events and entities.

The system is characterized by various "level convergence zones", all interconnected by feed-back and feed-forward relationships, which correspond to various levels of integration of the basic elements. According to the nature and complexity of the recalled memory, the convergence zones which are required for recall will not be the same. When an element of a situation already experienced enters the brain (via, for instance, one of the sensory channels), this very partial information reaching one (or several) zone(s) of convergence triggers, within the same time window, provokes the simultaneous activation of the areas that had formerly participated in the processing of the whole situation, by virtue of a combinatorial or binding code or addresses held in the convergence zone.

The arguments supporting Damasio's theory are derived from experimental neuropsychological studies (cf. Damasio, 1989b and 1990, for a detailed account of these arguments). A first set of data shows that anterior temporal cortex damage leaves intact the quality of perceptual experience, recognition, categorization, etc. This represents "generic memory". In contrast, the same patients are unable to recall the unique context in which some of these elements have occurred. Their "episodic" or "autobiographic" memory is therefore impaired.

Comparable deficits can be observed following anterior frontal lobe damage. An extensive ablation of the entire prefrontal cortices does not affect perception whereas drive for action, attention, problem solving, and evaluation of the recency of events are highly impaired. In contrast, damage to single-modality cortices deeply affects multimodal recognition and recall. For instance, a loss of color perception is associated with an impaired ability to imagine color in recall (Damasio, 1985; Farah, 1989). This supports the hypothesis that the same (primary and associative) sensory cortices are involved in both perception and recall.

The above summarized observations lead Damasio to formulate the following hypotheses. First, the processing of generic incoming information is completed well before this information reaches high level associative structures. Second, the storage of this information is ensured by these same areas which have participated in the initial processing, without the supervision of high level associative regions. Third, this information can be recalled without the contribution of high level associative regions provided that it is unambiguous and does not need to be inscribed within a unique autobiographical context.

In addition, the model assumes that there is not one unique convergence zone representing the culminating point of the whole process, but several, such as the prefrontal and parietal areas, for instance.

Damasio makes a distinction between three levels in the architecture of the brain ("early cortices", then "local convergence zones", and finally "non-local convergence zones"). Therefore, brain damage at a level of convergence which has not been involved in the global perception of a given situation or stimulus, i.e., which is located downstream from the required convergence zone, will not disrupt its recall. This is why the "earlier" in information processing is the cortical region damaged, the more likely it is that recall be affected. This explains the sparing of identification and recognition of generic objects and events which is observed when higher order associative regions are damaged.

In order to account for the formation of the reciprocal loops between various level convergence zones, Damasio proposes that Hebb-like mechanisms operate to strengthen the activated interconnections. For instance, the entorhinal cortex and hippocampus are brain regions where signals originating from many neocortical and subcortical structures co-occur over the same neuron ensemble. This feature makes the entorhinal area and the hippocampus good candidates for detecting and storing temporal coincidences.

As emphasized above, one of the original features of Damasio's theory is the taking into account of various level convergence zones. This appears more biologically plausible than dualistic views which envisage the brain as the hippocampus, on the one hand, and all the other structures on the other. Another interesting feature is that, as in Miller's model, emphasis is placed on the context and detection of ambiguities in perception and recall. Moreover, the notion of hierarchy of convergence zones leads to interesting predictions that could be tested in animals, about the impact of damage to early processing regions. Indeed, the only major drawback of the theory, which is common to other models, is that neither the nature nor the neural substrate of the binding code held in the convergence zones is specified.

Damasio's theory may be applied to spatial processing according to the following sketch: elements of a given representation (resulting from experienced events), such as local views, location representations, links between locations, conjunctions of cue configurations and movements, and so one, are separately stored in regions which have been involved in the specific processing of each of them. Several levels and structures (early cortices, local convergence zones) are involved in such fragmented storage. Higher associative regions (hippocampus and parietal cortex) would ensure the general organization of these scattered elements both during acquisition and recall. The detection of temporal and spatial coincidences among all of these elements would result in the formation of a combinatorial code for the representation of one given environment. Such a code may be located in the hippocampus. It may be further speculated that more abstract combinatorial codes, generalizable to various types of environment which share the same spatial properties, are held by other high level convergence zones, such as the parietal cortex. For recall to occur, the hippocampal combinatorial code should simultaneously co-activate the various elements of the composite representation so as to form a multimodal but unified representation whereas the parietal (and/or perhaps prefrontal) combinatorial code(s) may ensure computations and planning of trajectories. In this conceptual context, however, one would expect that damage to "early cortices" such as the occipital region would have much more deleterious effects than those usually observed in spatial tasks (Chapter 8), unless there is a full compensation for the loss of one sensory modality by others.

9. Complementary Learning Systems in the Hippocampus and Neocortex: McClelland, McNaughton, and O'Reilly's Model

The above authors have recently (1995) provided a model for learning and memory which accounts for many data and reconciles points of view which were hitherto divergent. The model is based mainly upon data showing, on the one hand, temporally graded retrograde amnesia following hippocampal damage and, on the other, the long duration of consolidation, which can be up to 15 years or more in some cases in human subjects. By examining the time course of the respective involvement of the hippocampus and neocortical areas in learning and memory, the authors sketch a picture of their interactions.

The model rests upon a large body of studies which have documented the phenomenon of progressive retrograde amnesia, reciprocal connections between hippocampus and neocortex, and on connectionist models and simulations (see McClelland et al., 1995, for a full account).

The authors assume that there are two complementary learning systems in the brain. The neocortical processing system relies "... on adaptation of synaptic connections among the neurons directly responsible for information processing and behavior" (p. 422). The distinctive features of this system (in the context of the model) can be summarized as follows: (1) many repetitions of an experience are necessary for the synaptic connections to be stable enough for the neocortical processing system to control the execution of a learned task; thus, the setting up of the memory trace is long and progressive. (2) The representation of an experience is widely distributed across the neocortical system, and (3) it may be very abstract (see below, for further comments on the notion of "structure").

The hippocampal memory system also relies on substantial modifications of the strengths of connections among hippocampal neurons following the transmission, via the entorhinal cortex, of inputs from the neocortical mantle, though (1) these modifications take place much more rapidly than in the neocortical system; (2) the stored information concerns specific episodes and events, in the form of conjunctions of inputs that are active in particular experiences; and (3) patterns of neocortical activation are represented in the hippocampus in a "summary sketch" (McNaughton, 1989b) or a "compressed" format which involves a much smaller number of neurons than do neocortical representations.

After having defined the respective properties of the two memory systems, McClelland et al. consider the reinstatement (restoration) and consolidation of hippocampal memories in the neocortical system.

Let us consider what happens of the memory trace of a particular (episodic) event (a learning session, for instance) stored in the hippocampal system as a stable memory and in the neocortical system as a weak pattern of activation. A first reinstatement of the hippocampal memory trace is likely to occur during rest or slow wave sleep phases. A second reinstatement will occur when the subject is subsequently exposed to the same learning situation which needs the memory trace for task performance. The authors assume that each reinstatement of the hippocampal memory trace
"... *would provide the opportunity for an incremental adjustment of neocortical connections, thereby allowing memories initially dependent on the hippocampal system gradually to become independent of it. To the extent that the hippocampal memory system participates in this reinstatement process, it can be viewed not just as a memory store but as the teacher of the neocortical processing.*" (p. 424.)

This mechanism accounts for the progressive incorporation of hippocampal memories in the neocortical system and for the corresponding decay of the hippocampal memory trace.

However, several other features of the model remain to be explained. The authors pose the following two questions: Why is a hippocampal system necessary, since ultimately memory traces are stored in the neocortex? And why does the incorporation of new information take so much time in the neocortex?

The responses provided by the authors to these questions are based on connectionist models and simulations. Briefly, the response to the first question is that the hippocampus exercises a function at the early stage of acquisition that precludes *interference* with memories already stored in the neocortical system. The response to the second question is related to that to the first: if the incorporation of hippocampal memories in the neocortical system were made rapidly these would interfere with the system of "structured knowledge" built up from prior experience with other related material. The presentation of new information has to be added gradually, *interleaved* with ongoing exposure to other domains of knowledge. Thus, the memory trace momentarily stored in the hippocampal system is protected from the risk of interference with the "structured knowledge" stored in the neocortical system. More importantly, this transient storage allows, in the meantime, comparison of that new information with structured knowledge so that it is integrated not only as an episodic event but also as an element which belongs to a "structure" held in the neocortical system, and/or participates in the elaboration of a new structure.

The notion of "structure" is central to the model. It is defined as follows by the authors:

"*What we mean by the term structure is any systematic relationship that exists within or between the events that, if discovered, could then serve as a basis for efficient representation of novel events or for appropriate responses to novel input.*" (McClelland et al., p. 436.)

This explains why the nature of information which is stored in the neocortex is different (more abstract and general) than that initially stored in the hippocampal system.

The model has many important implications. It accounts, among other things, for temporally graded retrograde amnesia following hippocampal damage. Other models, presented in the preceding sections, also accounted for this phenomenon but using quite different conceptual frameworks. It also provides original insights into the sources of variation in hippocampal decay and neocortical learning rates. More generally, it outlines the respective participation of the hippocampal and neocortical systems in the formation of categories and concepts (structures). In spite of the prominent interest of these issues, I will focus here only on the relevance of the model for spatial processing. In this respect, McClelland et al., go back to and develop their initial interpretation of the properties of hippocampal neurons (see section 4), namely, that they appear to encode conjunctions of cues (including the animal's inertial sense of direction), locations and movements. In other words, the hippocampal system would facilitate the linking of experiences to their spatial contexts. In the longer term, it would prevent interference occurring among various spatial representations, but the authors do not speculate about the role of the neocortical memory system in spatial processing. It appears, however, that the notion of structure refers to cognitive maps insofar as one admits that the properties of such high-level spatial representations are not specific to one environment but can be generalized to various situations. This issue is developed in Chapter 10.

More specifically, McClelland et al.'s model sheds light on the puzzling data obtained by Poucet et al., (1991a; see Chapter 6): temporary inactivation of the hippocampus (during learning sessions) did not interfere with distributed-trial spatial learning but prevented any kind of acquisition with a massed-trial procedure. Training sessions with the distributed-trial procedure were interleaved with a 23.5 hour periods of rest, sleep and usual activities in the home cage. However, the fact that during subsequent training sessions, with the hippocampus inactivated again, animals were still able to make efficient use of spatial information to reach the goal location remains unexplained.

10. Concluding Remarks

The first models described in this chapter emphasize various levels of complementary spatial processing which may be implemented either by the hippocampus alone or via cooperation of the latter with neocortical areas. The other models concerning more general aspects of memory provide hypotheses regarding the possible nature of such cooperation. One of the aims of this brief overview has been to stress the complementarities rather than the differences that may exist among various conceptual frameworks, whether they specifically deal with spatial processing or not. Indeed, as illustrated by the forewords, the fact that very different aspects of the contributions of brain structures to spatial processing may be stressed by the various accounts, does not necessarily mean that they are mutually exclusive.

Let us first consider the models which specifically deal with spatial processing. The basic level, which is the building block of most of the models, relies on the firing properties of hippocampal place cells. In this respect, Sharp provides an account of the setting up of these properties which is in agreement with behavioral data, in particular with those demonstrating the spatial function of exploration. Multiple and overlapping local views available from the same place, and of the same layout, are necessary for spatial knowledge to be achieved. Muller and coworkers' model for connectivity encoding is quite compatible with Sharp's proposal. Both accounts have been incorporated by Poucet into a broader view of spatial processing.

In contrast, for both O'Keefe and McNaughton (McNaughton et al., 1993a), place cells encode environmental landmarks in terms of vectors. Each place cell represent a vector which is specific to one (or two) landmark(s) for McNaughton et al. or which is the vector average of a larger subset of cues (O'Keefe).

On the basis of different mechanisms, Sharp and O'Keefe account for the same phenomenon, i.e., how place cell ensure the integration of egocentered perceived visual scenes or landmarks into an exocentered framework; thus how place-encoding becomes independent from the animal's head direction but specifies its current location within the absolute space. In contrast, place cell vector encoding as proposed by McNaughton et al. (1993a) defines the animal's position (head direction) by reference to one landmark. The observation of place cells with strong directional correlates in the radial maze provides some support for the latter hypothesis.

Another feature common to several models and which is straightforward to interpret is the function of head-direction cells conceived as supplying to the system an overall reference direction. The latter may result from the computation of a slope (O'Keefe), or head-direction cells may serve to maintain an "inertial direction sense", or as a bearing for vector encoding by place-cells (McNaughton and coworkers), or finally, they may provide a full range of reference directions for various location-dependent frameworks (Poucet). By the same token, in all cases, movement-related inputs (from the parietal cortex, except in O'Keefe's model which does not explicitly refer to the origin of this type of information) are used to update the animal's current heading by reference to the overall direction. Thus, the models of spatial information processing which have been presented here appear either complementary or at least to rely on more or less identical principles, with regard to the earlier stages of the process and/or basic elements such as place and overall direction encoding. In contrast, there are divergences when the combination and use of these basic elements for computing trajectories and navigation are considered. At this level, experimental arguments to support the various hypotheses are sparse. Accordingly, the authors are led to speculate much more than when a consistent set of data is available, as is the case for place cell and head direction cell properties. Each model has its own originality which has been outlined above. The proposals which rely on relatively simple mechanisms, such as the two proposed by McNaughton, can be viewed as means of orienting that do not preclude the implementation of more elaborated representations, and consequently are not exclusive of other scenarios. In contrast, O'Keefe's and Poucet's models, which account for navigation based on representations of similar cognitive levels, have fundamental conceptual dissimilarities. Among others, O'Keefe assumes that animals construct and use one unique form of spatial representation, metric maps, whether extended or limited regions of space are concerned. On the other hand, Poucet envisages a progression of the complexity, and levels of abstraction, of the structure of representations (which first have topological properties, but later metric properties). However, it is obvious that the predominant distinctive feature of O'Keefe's (and Nadel's, 1991) theory, by comparison with the models advanced by Poucet and by many other authors, is that the hippocampal formation (1) mediates all spatial processing, and (2) continues to perform its same mapping function during the early and late stages of processing.

The first point seems to be partly a matter of semantics. If the definition of spatial processing is restricted to the mapping process, O'Keefe and Nadel's assertion is not totally at odds with other conceptual frameworks in which hippocampus is assigned a major spatial function. Yet, as has been discussed in Chapter 7, the parietal cortex seems to play an important role in the transformation of egocentered information into an allocentered frame of reference (and perhaps in the reverse transformation as well) by providing, among other things, movement-related feedback. This might correspond, in O'Keefe's theory, to the preprocessed sensory information which reaches the mapping system via the entorhinal cortex. In contrast, the involvement of the prefrontal cortex in the "translation" of spatial representation into action is not considered at all by O'Keefe, though it is true that this aspect of spatial processing does not concern mapping mechanisms per se but rather the use of the resulting map. Thus, apparent contradictions between O'Keefe's theory and other models seem to arise in part from variation in (narrow versus broad) definitions of spatial processing.

The second distinctive feature of O'Keefe and Nadel's theory is the long-term persisting role assigned to the hippocampus in spatial memory. In this respect, Nadel (1991) wrote about the sparing of spatial performance following hippocampal lesions when these were made a certain time after learning:

"*What is always overlooked in these discussions is the simple logical fact that such demonstrations say nothing about the status of memory storage in the hippocampus – they address only the question of whether there is adequate information outside the hippocampus to support performance. It is very interesting that the establishment of such adequate extrahippocampal storage takes time, but it does not logically follow that during the same time period information is disappearing from the hippocampal system.*" (Nadel, 1991.)

The implacable logic of the argument cuts further discussion short, awaiting evidence of the decay or persistence of spatial memory storage in the hippocampus. What is uncontroversial is that information is also stored in neocortical (and subcortical) areas and that after a sufficient post-training delay, this information can ensure spatial performance in the absence of the hippocampus and/or related structures.

The first three general models of memory (Teyler and DiScenna, Miller, and Damasio) have been presented as illustrations of the possible functional interactions between the hippocampal formation and other neocortical and subcortical structures. All three rest on the same general principle, though implemented via different neural mechanisms. Schematically, the main idea is that the hippocampus (and other high level associative regions, for Damasio) holds some "code" or "addresses" of the various brain areas which have come into play during the actual (then, later, remembered) experience. Such codes allow for simultaneous co-activation of these brain regions during recall. This implies that memories are not stored as a whole but in a fragmentary form, distributed into many brain regions. Another more or less implicit implication is that the hippocampus plays the role of a "conductor", simultaneously entraining or regulating several loops which are active in parallel. The common anatomical substrate for the three models is the large number of direct and indirect reciprocal connections which link the hippocampus (and other associative areas such as the frontal and parietal areas) with many other neocortical and subcortical brain regions. After consolidation, connections among the various co-activated areas would be established so that the hippocampus would not be longer necessary.

These models are relevant for spatial memory, as for any other type of memory. In addition, they lead to consider the possibility that a spatial representation may not be stored as a whole, but rather in the form of distinct constituent elements (see Worden, 1992, "Navigation by fragment fitting", for a discussion about the advantages of such a hypothesis).

The last theoretical framework, that proposed by McClelland et al. (1995) accounts for the progressive incorporation of memories initially formed in the hippocampal system, into neocortical areas. The authors consider their model to be "synergistic" with other accounts (such as that advanced by Teyler and

DiScenna, and Damasio; see McClelland et al., 1995, for a discussion of this issue) instead of being mutually exclusive. They stress that McNaughton's (1989b) notion of a hippocampal "summary sketch" or "compressed" representations in the hippocampus which serve for reconstructing full representations in the neocortex is very germane to that of a hippocampal code or index found in the other theories. Yet, McClelland et al.'s assumption appears to be more plausible at the biological level than that advanced in the other proposals since the authors explicitly mention that hippocampal neurons represent "specific combinations or conjunctions of elements of the event that gave rise to the pattern of activation" (McClelland et al., 1995, p. 424).[c] Though not explicitly stated by the authors, this would mean that the events are not stored in the hippocampus. In this hypothesis, the hippocampus would mainly store a compressed representations of such events.

Lastly, the model accounts for many existing data and leads to important predictions about memory. Moreover, one of its properties is of major interest for better understanding spatial processing in that it accounts for the transformation of episodic, situation-specific (spatial) representations into more abstract and generalizable "structures".

---

[c] This view is also very close to Sutherland and Rudy's (1989) hippocampal "Configural Association System". In a recent appraisal of their theory, Rudy and Sutherland (1995) propose that "... the hippocampal formation contributes to configural processing by enhancing the activation or salience of configural or conjunctive representations constructed in other areas of the cortex" (p. 382).

CHAPTER 10

CONCLUSIONS

"Cognitive maps are often discussed as if they were mental pictures of the environment that could be examined at leisure by the mind's eye while the mind's owner reclined in his armchair." (Neisser, 1976, p. 110)

I drew attention in the first chapter to the fact that the term "cognitive map" could be misleading because it conveys the notion of a static entity. Throughout the present book, the majority of questions that have been raised concern the interpretation of data in relation to the construction and use of cognitive spatial maps: to what extent do such-and-such results provide evidence that animals possess cognitive maps? Are there more parsimonious hypotheses to account for the observed behaviors? And so on. Some of these questions arise from the fact that the "cognitive map" notion is seldom taken as a metaphor, but more often stricto sensu. Loosely defined by Tolman, often unwittingly used by many authors, although more precisely circumscribed by O'Keefe and Nadel (1978), the term is not satisfactory. A first imperative should be to propose an alternative term for high level spatial representations in animals. I would propose the term "spatial integrating system" as a suitable replacement, since this term emphasizes an aspect which had been stressed by Tolman but more or less neglected by present day researchers, namely the dynamic aspect of spatial representations. In my view, taking into account this property of so-called cognitive maps may assist the better understanding in the future of their underlying nature and properties, and the setting up of new experimental hypotheses.

In the first section of the present chapter, I explain why high level spatial representations should be viewed as active structures for organizing new incoming information. Then, I present a multi-stage sketch of spatial processing based on the dynamic feature of the spatial integrating system. The following three sections are devoted to comments about this dynamic view, in relation to some of the important issues that have been discussed in the preceding chapters and that appear to be of particular relevance for further research. Lastly, the main questions that are raised relating to the implementation by the brain of such an sketch of spatial processing are discussed, followed by my concluding comments.

1. The Spatial Integrating System: an "Active Information Seeking Structure"

Representations, whether they are spatial or non-spatial, map-like or not, are entities, the nature of which can only be deduced from observation. The most classical (i.e., most frequently used) means of investigating their nature is to measure animals' performances in traditional training and problem-solving tasks. Though such a method has proved to be extremely useful, it does not say anything about *how* animals reach (or fail to reach) high performance levels. In this respect, the study of the strategies implemented by animals for solving tasks allows us to gain further insight into the mechanisms which underlie the observed behavior. The analysis of patterns of visits in the radial arm maze, and the computation of parameters of rats' trajectories in the Morris swimming pool task are examples of this type of analysis. Another approach which is instructive as to the dynamic aspects of spatial processing is the otservation of exploratory activity and reaction to spatial changes following habituation. These "dishabituation" paradigms can yield some of the spatial parameters that are spontaneously encoded by animals while exploring. At a more general level, such studies demonstrate that spatial knowledge involves active processing which relies on multisensory (mostly visual) perception of the environment matched with movement-related information. However, one question, related to the control of exploratory behavior, has never been addressed in animals, namely, is exploratory activity randomly organized, or is it controlled by any form of pre-existing cognitive "structure" or "schema", that functions to optimize information gathering? In this respect, it is significant that Neisser (1976) had advanced an interesting definition of a cognitive map that he conceived as an "orienting schema" or "active information seeking structure" which directs both exploration and action.

Data generated from dishabituation experiments can be interpreted in Neisser's terms. Re-exploration of a spatially modified situation appears to correspond to an updating of the initial representation (Thinus-Blanc, 1987) that would control the renewal of exploratory activity, on the basis of a detected mismatch between the initial, represented configuration and the currently perceived configuration. Since exploratory activity can be observed, the analysis of behavior patterns should reveal some properties of the unobservable structure that directs and organizes them.

Several attempts have been made in my group (in yet unpublished studies) in order to identify regularities from recorded sequences of contacts made by animals with objects scattered in an open-field, though we have failed to demonstrate any emerging patterns. Rather, randomness appeared to be the main "organizing" factor for exploration, at least, as we had quantified exploration, i.e., in terms of contacts with the objects. We concluded that the multiple back-and-forth displacements might have an auxiliary function in preventing the repetition of the same patterns of stimulation likely to result in the learning of sequences of stimulus views or local views.

More recently, we have been addressing a question that, at first glance, has no direct link to the above issues, that of understanding the nature of spatial deficits that are often observed in congenitally blind persons. The available literature in this domain is rather confusing. On the one hand, some studies show that early or congenitally blind subjects are severely impaired in spatial processing, by comparison with late blind and sighted blindfolded subjects. On the other hand, no differences among the three groups of subjects have been found by some authors in situations that are quite similar to those in which impairments have been reliably observed by others (see Thinus-Blanc and Gaunet, 1996, for a review). One possible explanation for the divergent data may be related to the selection of congenitally blind subjects. A diverse range of orientation and mobility training methods are used with blind children, and they interact with many individual features and past history (family environment, affective reactions to blindness, and so on). Such factors are difficult, if not impossible, to control. Therefore, in order to better understand the origin of the inconsistent pattern of data, we have analyzed behavior occurring "upstream" from task achievement, i.e., at the moment when spatial information is acquired. The purpose was to examine whether differentiated performance levels could be correlated with differentiated means of collecting spatial information.

Two reaction-to-change experiments, based on the paradigms used in animal studies presented in Chapter 2, have been adapted for use with human subjects. One of them took place in a room which the participants had to explore by walking round a configuration of easily identifiable objects (the "locomotion space"; Gaunet and Thinus-Blanc, 1996). In the other experiment, the subject was seated in front of a tray on which a set of objects was displayed, and they explored the objects with one hand (the "manipulation space", Gaunet, Martinez, and Thinus-Blanc, 1996). Following a fixed period of exploration,[a] a modification was made to the initial arrangement and participants were asked to explore again until they detected whether a change had been made. During the pre-test exploration phases, the sequences of contacts that subjects made with the objects were carefully recorded and analyzed. Two types of pattern were found to stem from exploratory sequences in both locomotion and manipulation spaces: One of them, termed the "Cyclic pattern", consists of visiting all the objects succes-

---

[a] Habituation was not taken into account. Preliminary observations had shown that the duration of exploration (15 mn and 3 mn for the locomotion and manipulation situations, respectively) had indicated that this duration was sufficient for a representative control subject to reach a high performance level.

sively, beginning and finishing at the same one. The other pattern – the "Back and Forth pattern" – involves making repeated contacts between two objects (Fig. 10.1). Both types of exploratory pattern have been found to be implemented by all the three populations of participants (congenitally blind, late blind, and blindfolded sighted subjects). But how did strategy use relate to subjects' performance levels? Results showed that congenitally blind subjects were impaired at detecting the change. The interesting point is that the early blind subjects made many more "Cyclic" than "Back and Forth" exploratory patterns whereas the reverse was observed in the other two groups. Statistical analyses revealed performance level to be correlated with the proportion of contacts involved in each type of exploratory pattern; the higher the performance level, the smaller the proportion of contacts that were of the "Cyclic" pattern type, and vice-versa.

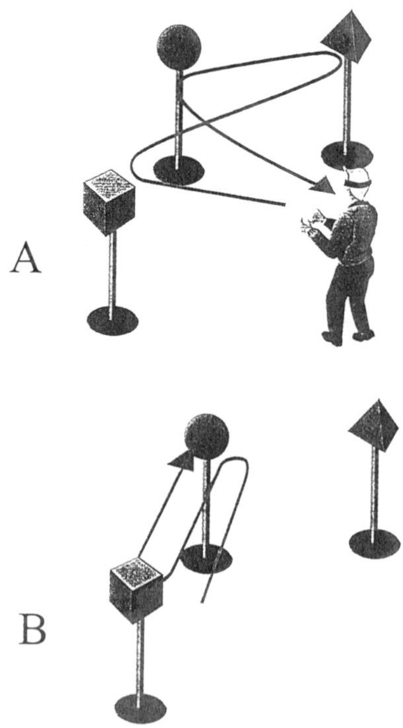

Fig. 10.1. Example of "Cycle" (A) and "Back and Forth" (B) patterns of exploration. (Adapted from Thinus-Blanc and Gaunet, 1996.)

Though they raise many questions, related, among other things, to the meaning of Cyclic and Back and Forth patterns, these data support the idea that, at least in humans, some organizing strategies underlie spontaneous exploratory behaviors, which result in more or less accurate spatial representations. In addition, they suggest that our failure to identify any regularities in hamsters' exploratory activity may have been due to the fact that, unlike subjects in the human experiments, animals could pick up distant visual information (indeed, many stops with scanning head movements were frequently observed). Therefore, sequences of actual contacts with objects may not have fully reflected the organization of information gathering.

Lastly, preliminary (unpublished) results obtained in rats suggest that exploratory activity (strategies) may be controlled by whether or not animals have the opportunity to generalize spatial knowledge to different situations which share basic geometrical features. Rate of habituation was examined in two different situations, over twelve four-minute sessions of exposure. In the former situation, different objects placed in a circular open-field defined an irregular geometrical shape which remained the *same* across all sessions, but the position of each of the four objects changed from one session to the next. Thus the shape of the arrangement was permanent but the combination of the objects defining it was variable. In the other situation, the same four objects were used but they defined a *different* geometrical arrangement on each session. The aim of the experiment was to examine whether rats are able to extract the structure common to different situations. In such a case, the rate of habituation should have been more pronounced when the general structure remained stable over sessions. As expected, repeated exposures to the same geometrical configuration resulted in a higher rate of habituation, in terms of number and duration of contacts (Fig. 10.2).

At this stage, neither rat nor human studies provide explanations as to the mechanisms which underlie the results described above. Yet they provide some grounds for stating, as a working hypothesis, that exploration depends on, and is organized according to, representations pre-existing in advance of exposure to the current situation, or progressively constructed during repeated exposure to that situation. Data obtained in the experiment with rats support the idea (one that has already been advanced in Chapter 2, based on the lack of re-investigation when the square defined by four objects is decreased or enlarged) that animals appear to form abstract representations of spatial configurations sharing common spatial features. The term "abstract" is used here and throughout the present chapter in its broad meaning: it does not necessarily refer to the highest level of abstraction but it applies as soon as there is a generalization process taking place. It refers to any cognitive processing, the result of which is not bound to one unique feature or set of features of a given environment but which can be generalized to various other situations.

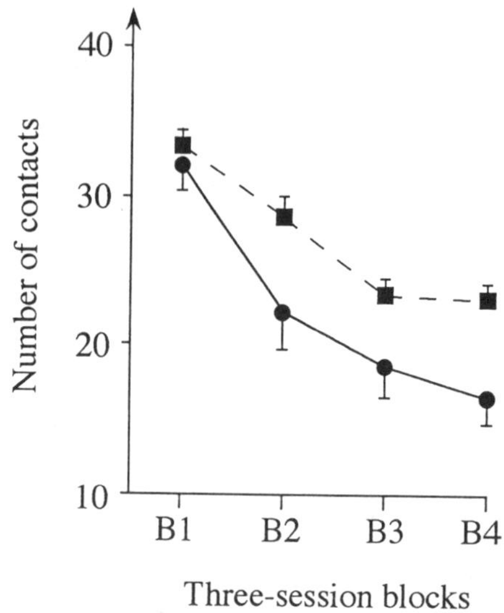

Fig. 10.2. Evolution of habituation (numbers of contacts) over the three-session blocks of the experiment for the groups with a constant (circles) or variable (squares) object arrangement. N = 8 in each group. (From Thinus-Blanc, Save, and Poucet, unpublished study.)

2. A Dynamic Sketch of Spatial Processing Mechanisms

Based on the data presented above, and along the lines of the hypothesis initially advanced by Neisser (1976), a dynamic view of spatial information processing can be portrayed as in Fig. 10.3. The process takes place at two distinct levels: that of action within the physical world, and that of spatial representations. The first level corresponds to observable behaviors (exploration, oriented trajectories) and to information (local views and movement-generated feedback cues) that can be controlled by manipulating some features of the situation. By its very nature, behavior is organized along an egocentric reference frame (e.g., go ahead, turn left) and sequentially ordered (go ahead, then turn left) even if this sequence and the bearing of the goal have been determined on the basis of exocentered representations (see later). In contrast, a map-like representation holds allocentric information, and its constituent elements are charted regardless of the sequence of actions which took place in the course of their acquisition.

As has been discussed in preceding chapters (e.g., Chapters 2 and 7), external (local views) and internal (movement-related) inputs generated during exploration have to be matched so as to extract spatial invariants. This stage corresponds to the interface between the level of action and that of representations. Spatial invariants extracted from a given environment are the constituent elements of the representation which is *specific* to that environment. This represen-

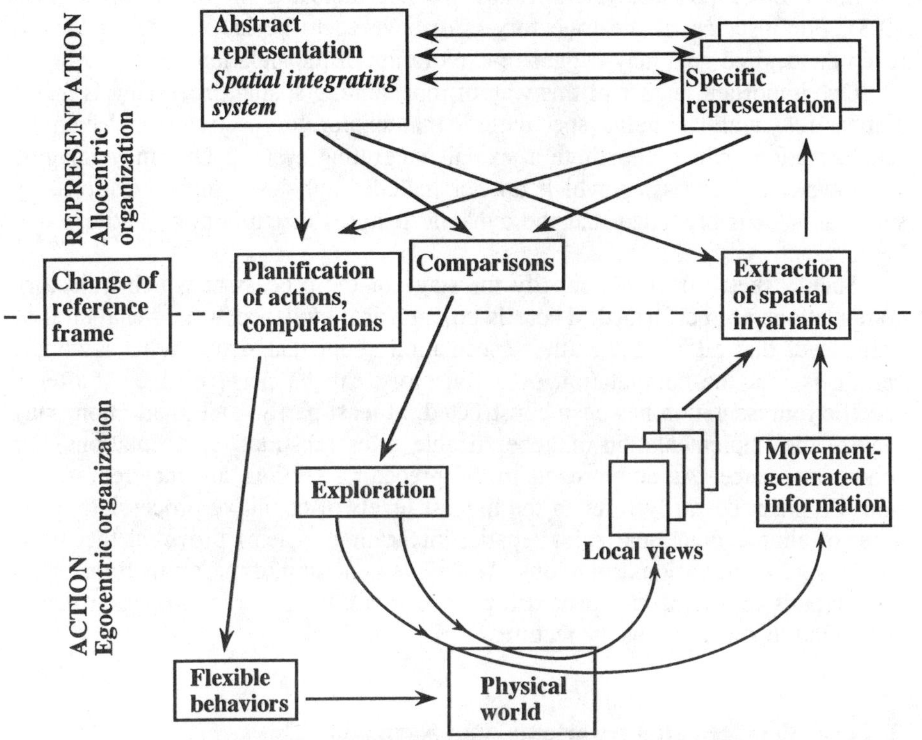

Fig. 10.3. A sketch of the spatial integrating system (see text).

tation has several functions. First, it may be compared to current inputs resulting from the detection of a change brought about to the initial situation. The detection of a mismatch between the stored representation and the sampled modified spatial layout will reactivate exploration in order to update the existing representation. Another role of specific representations is their contribution to the formation of more *abstract* representations (or "structures" according to McClelland et al.'s [1995] model; see Chapter 9) holding general spatial properties common to several specific representations. Finally, together with abstract spatial representations, specific representations allow for the computation and planning of oriented trajectories within the physical world. The latter process is located, like the system for extracting spatial invariants, at the interface between representation and action, but it operates in the reverse direction. It is assumed to translate an exocentered representation into the form of actions to be achieved along a temporally ordered sequence and organized in an egocentric framework.

Some other possible relationships are not indicated in the schemas (Fig. 10.3). For instance, if the trajectory is discovered to be inadequate, failure to reach the desired goal may trigger re-exploration of the situation.

One important feature of this way of representing spatial processing is that it emphasizes, without undue speculation, that exploratory activity is driven by representations which constitute a spatial integrating system. One may imagine other types of processing which do not include high level representations. In such cases, it is predicted that the dynamic features of exploratory activity will be different.

Such a sketch does not specify the stage of the process at which trajectory computations are performed. There is no empirical evidence at the behavioral or brain level that allows for further speculation about that issue. What is almost certain is that the computation of a trajectory can be performed only after a specific representation has been constructed, at least partly, and in addition, may require the implementation of generalizable rules (abstract representations) for processing space. As emphasized in the preceding section, abstract representations do not necessarily refer to the highest levels of cognitive processing but to a set of simple functional rules (spatial integrating system) provided that these can be applied to various situations. The implementation of these functional rules corresponds to an abstract processing since it results in a representation that is not bound to specific sensory features.

3. Local Views, Spatial Invariants, and Navigation

As underlined several times in the present book, spatial representations, whether they are map-like or not, result from the perception of the environment, which is necessarily dependent upon the subject's position and is sequentially ordered in time. Even though the visual flow while moving is a continuous stimulation, one of the consequences of visual scanning, making head movements that enlarge the zone of inspection, and locomotor activity is the generating of different visual scenes. These scenes have been called "local views" since they depend on a given position of the eyes, head and body at any given moment. These are the building blocks for further spatial processing.

Most scenes which are generated in the course of the activities of everyday life are stored only in short-term memory. Those which are stored in long-term memory are supposed to be of special importance because they may prove to be necessary for further use in the future. For instance, local views corresponding to a marked change in the perceived spatial layout, such as those seen when turning a corner or reaching a cross-road in a maze, may be stored in long-term memory. The number of local views required to correctly "label" an environment is a function of the complexity of that environment.

Obviously, the visual modality is not the exclusive means of acquiring spatial knowledge. However, visual information is much more accessible from afar, more precisely topographic and more manipulable than information provided by any other perceptual modality. Vision also provides simultaneous stimulation from multiple sources. It is a prominent sensory channel for gathering and organizing spatial information (Thinus-Blanc and Gaunet, 1996).

Local views correspond to a *concrete* level of spatial description since they contain information that is tightly related to a direction of approach or a vantage point. Therefore, if a location is specified by one (or several) local view(s), its use for navigation is likely to be bound to the direction(s) from which it is approached. When locations are specified in terms of local views, that leads to the implementation of routes (O'Keefe and Nadel, 1978; see Chapter 1) or variants of routes such as S-R-S associations (McNaughton and coworkers' Conditional Association model, see Chapters 5 and 9). Each trajectory between two places depends upon the immediately-preceding movement which leads the animal into the proper perspective for recognizing the local view that best matches the stored one. For spatial representations to be flexible, they need to be freed from the constraints related to appearance; it is necessary that places, and angular and distance relations among places are processed in a more abstract way.

As a matter of fact, the term "place" can refer to various kinds of entity. According to a strict geometric definition, a place is a locus in the sense of being a point. In vernacular usage, a place may instead refer to a zone that has extent and that contains one or more elements or objects. In the present context, there is another important distinction: that between places occupied by the subject, and places directly defined by particular features (such as the goal in a maze, a set of objects clustered in a field) or indirectly defined on the basis of environmental landmarks (because they are empty and featureless by themselves).

The abstract processing of the place occupied by the subject her or himself can be accounted for in relatively simple terms (Kuipers, 1982; Poucet, 1993; Thinus-Blanc, 1992, 1995). The various overlapping local views provided by a 360° rotation on the spot have in common the point from which they are perceived (Fig. 10.4). In that case, a place may be abstractly defined as the *federating core* of panoramic local views. Sharp (1991; see Chapter 6) has provided a valuable model of how this type of processing could result in the setting up of hippocampal place-cells that fire regardless of the subject's head direction but at a given location. A number of findings (Chapter 6) and theoretical models (Chapter 9; O'Keefe's model, in particular) support the idea that a subject's location encoding is made by reference to stable environmental landmarks. Such places have a transient specific status since they are related to the observer's location. When the animal moves, its' changes of location have to be updated by reference to the environment.

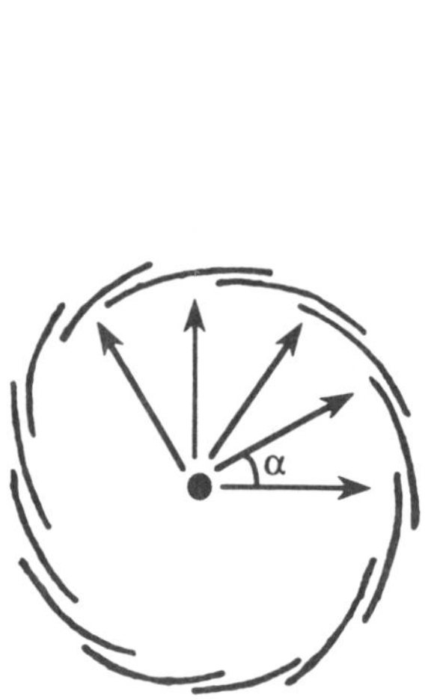

Fig. 10.4. Illustration of the definition of a place based on the overlapping local views provided by a rotation on site. (Reproduced from Thinus-Blanc, 1995.)

Another type of processing which is complementary (and has reciprocal relationships with the first) refers to the notion of place and spatial relationship invariance. In the same way as there is a system that defines the animal's location regardless of the perception of the environment for a given head-direction, there must be processing of places, and more importantly of relationships among places, which is also independent of the subject's vantage point. When a place is labelled by an object, its invariance is related to the notion of "object constancy" (see Chapter 5). Usually, complex spatial problems imply the computation of the location of a hidden goal on the basis of perceived landmarks. Although Collett et al. (1986) argue that only one landmark is used to compute the trajectory to an inconspicuous goal, other data (e.g., Suzuki et al., 1980; see Chapter 3) support the idea that spatial relationships among landmarks are used to deduce the goal location. Such relationships must be identified. It is legitimate to use the term "abstract" to describe the result of place and spatial relationship processing, since the derived knowledge, though based on a concrete description, becomes independent of it.

Some data, though sparse, suggest that spatial orientation can be accurate even from a totally novel (previously unexperienced) direction of approach (Morris, 1981; Olton, 1979). In that case, the local view and relationships among landmarks perceived from an entirely unfamiliar perspective may not be recognized. Yet, it is necessary that the animal keeps heading toward the bearing, the computation of which has to rely on a firm basis – sticking to a firm course – since there are no "reassuring" cues. Information provided by an overall reference direction must prevail in such a case over any other type of information.

In experimental settings, which are usually of restricted size, it is unlikely that animals will be confronted with totally unfamiliar perspectives of the landmark array, unless steps have been deliberately taken to control this aspect by preventing animals from seeing some parts of the spatial layout (see Sutherland et al., 1987, and Chapter 3). In contrast, in large spaces, it is more likely that animals will encounter this type of problem since they may have to navigate within relatively unknown regions and to orient toward their home from unfamiliar directions of approach. In such cases, it can be assumed that the trajectory to the goal consists of a first segment roughly oriented toward the region of the goal, followed by a more precise ballistic adjustment on the basis of the familiar landmark array. In contrast to laboratory conditions, it is improbable that animals in natural conditions have an homogeneous knowledge of the whole domain in which they move around.

The above comments lead to examination, in the next section, of the possibility that different orienting mechanisms are implemented according to the size and complexity of the environment.

4. Navigating in Visually "Disconnected" Environments

Poucet (1993), in his model, made a distinction between spatial processing (and related navigational mechanisms) in restricted regions of space (local charts), and in larger environments which encompass a number of local charts (see Chapter 5). His definition of local charts is based on the idea that these "... may contain any number of place representations but, by definition, all of these place representations must share a number of common stimulus elements provided by either the proximal or distal environment". (Poucet, 1993, p. 170). In other words, the same landmark array can be perceived from both the origin of a trajectory and its goal. That is exactly what happens in most of the experimental situations used so far. I have expressed several times, in the preceding chapters, reservations about the interpretation of high performance levels recorded in such situations in terms of overall representations or cognitive maps. Indeed, more parsimonious explanations, such as the Vector Sum hypothesis, can account for successful behaviors. In contrast, a crucial test of the cognitive mapping hypothesis would be to demonstrate that animals are able to compute direct trajectories from a part of the environment which does *not* share common "stimulus elements" or landmarks with that of the goal.

In humans, this type of representation is in everyday use. We can compute the optimal path to be taken (considering the traffic density, one-way roads, and so on) for traversing a large city. Though there are no landmarks common to both the place from where we start and that of our final destination, they can be connected in our representation. Indeed, the demonstration of such a mechanism in animals would provide strong support for the cognitive map hypothesis by ruling out any other more parsimonious hypotheses. Using the Morris swimming pool task, Sutherland et al. (1987) addressed this issue by using partitions preventing rats either from swimming directly from one half of the pool to the other or from viewing the whole situation (Chapter 3). Rats which experienced the pool in this fragmented way were found to take longer to reach the escape platform.

Recently, Schenk and Gafner (in prep, reported by Schenk et al., 1995) designed an elegant experimental situation aimed at testing spatial abilities without visual information about environmental landmarks. Rats were allowed a full view of distant environmental cues from the periphery of a "homing board" (Fig. 10.5, left panel). When they were in the central region of the board, they were prevented from seeing these landmarks. The goal was a hole in the floor of the apparatus, which was connected to the home cage, under the board. The task consisted of escaping through one of the four holes at a fixed spatial position relative to the visually disconnected landmark array. Consequently, an accurate localization of the correct hole implied that animals could construct and use a representation including two visually disconnected environments: the bare arena with the goal location and the distant cue array. Training was followed by probe trials during which escape was not possible, in order to rule out the use of cues emanating from the goal hole. The amounts of time spent in the hole sectors (Fig. 10.5, right panel) were recorded. Though the poor performance of one of the groups (tested in the "free-access" condition; see legend of Fig. 10.5) is difficult to interpret, it is clear that rats are capable of locating an inconspicuous goal located in an environment visually disconnected from relevant spatial information. This study raises interesting issues that are very directly related to the question posed at the beginning of the section: for instance, could animals compute a new trajectory (shortcut or detour) within a situation which encompasses two visually disconnected environments, each of them being equally rich in visual cues? The situation used by Chapuis et al. (1987) (Fig. 4.5) to study shortcuts could be adapted to provide an answer to that question by visually isolating each of the two subspaces.

5. Abstract Representations and Derived Rules for Spatial Processing

In my opinion, another important issue raised by the notion of spatial integrating system is related to the possibility or otherwise of generalizing representations and extracting some invariant features from the experience of a number of different situations. Indeed, the demonstration of animals' capability of setting general rules for encoding spatial relationships and for computing accurate

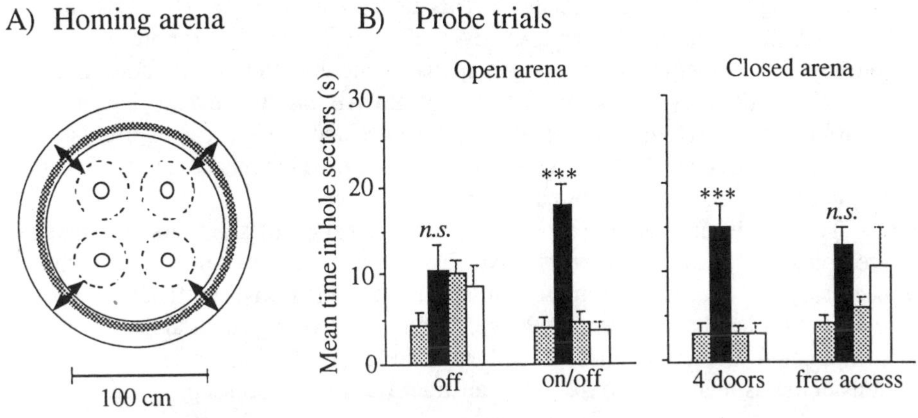

Fig. 10.5. A: The two-compartment table. In one situation (open arena), the inner region was defined by switching off the room lights each time the animals entered it ("on/off" group). A control group ("off") was trained in total darkness (under infrared light). In another condition (closed arena), the inner region was defined by a white curtain with fringes touching the surface of the table. In such a case, animals could access the arena from any direction by passing through the fringes ("free access" group). In another "closed arena" condition (represented on the Fig.), a wall defined the arena which could be accessed only through four doors ("4-door" group). B: Time spent in he hole sectors during the 72 s probe trial with no hole connected. A control group had been trained in constant darkness ("off") in the open arena. As expected, they did not show any preference for the region of the escape hole. In contrast, rats from the "on/off" group spent significantly more time around the escape hole. The same result was found in the closed arena condition for the "4-door" group. In contrast, confusion between holes was recorded when access was unrestricted through the fringes ("free-access" group). However, that group improved performance level up to that of the 4-door group with additional training. (Reproduced from Schenk et al., 1995).

trajectories, would be an additional powerful argument in support of the spatial integrating system.

In section 2, possible rules for extracting spatial invariants (place representations and relationships among places) have been described. It is also possible to account for functional relationships by the implementation of simple rules which may be inferred from experience.

Any configuration of at least three elements contained in space can be described as a triangle (a "mesh", Chapuis, 1988) and all the other arrangements, as combinations of triangles (Thinus-Blanc, 1992, 1995) forming a network. The trajectories of any organism between the three apexes of a triangle have several functional properties which constitute what Piaget (1937) called the "groupe logique des déplacements". This group can be summarized as follows (Fig. 10.6): (1) the routes are transitive: AB and BC can be combined into one journey AC belonging to the same spatial structure and which does not run through B; (2) they are reversible: any route AB has its reverse BA; (3) the combination of AB and BA results in no route; and (4) the routes are associative: in the sequence ABCD, AB+BD = AC+CD. This means that the location D can be reached by different pathways.

The two basic spatial computations, determination of shortcuts and detours, are contained in these few words, which summarize in functional terms what is necessary to successfully travel in space, provided that sufficient vectorial information has been acquired (Poucet, 1993). The rules which allow for vector computation along the lines of Collett et al.'s (1986) hypothesis can also be considered as belonging to the set of abstract rules for processing space.

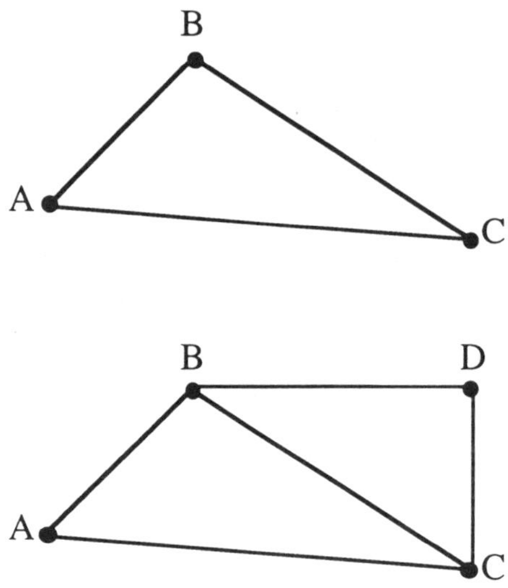

Transitivity: AB+BC=AC
Reversibility: AB implies BA
Annulation: AB+BA=0
Associativity: AB+BD=AC+CD

Fig. 10.6. Illustration of Piaget's "groupe logique des déplacements". (Reproduced from Thinus-Blanc, 1995.)

The experiments which have been conducted so far (Chapter 4) allow us to conclude that animals have stored such general rules but this has been demonstrated only in experimental situations of restricted size and complexity corresponding to Poucet's local chart definition.

6. Implementation of the Spatial Integrating System at the Brain Level

In this section, I consider how various brain structures can ensure the implementation of the multi-stage spatial processing outlined at the psychological level in Fig. 10.3. The function of some brain regions is unambiguous, or at least widely accepted. For instance, the implication of primary and secondary sensory cortices in the perception of local views (which may not be exclusively visual) is indisputable. The role of the associative parietal cortex in the extraction of spatial invariants appears also to be supported by the convergent multisensory (mostly visual) and proprioceptive inputs to this area, though some integration may occur upstream (for instance, vestibular nuclei receive both vestibular and visual inputs). Similarly, the working memory function of the prefrontal cortex has been demonstrated in many studies; the assertion that it would hold in short-term memory the information necessary to plan and execute a trajectory is hardly speculative. Lastly, among the undebated issues, the conception of the hippocampus by O'Keefe and Nadel (1978) in terms of a comparator or detector of mismatches between stored representations and currently perceived situations has not raised serious objections.

In contrast, the implementation by the brain of other spatial processing stages is still hypothetical. For instance, the calibrating role of vision in integrating non visual information into a visual canvas, and the storage of local views[b] in the occipital areas which have mediated their initial processing is proposed as a working hypothesis on the basis of data presented in Chapter 8.

Concerning the level of interface between representation and action, I have suggested (Chapter 7) that the associative parietal cortex, besides the transformation of perceptual information into representations, would achieve a similar transformation in the reverse direction, from representation to action, specifying, for instance, as motor commands directions to be taken. The only experimental support for such an hypothesis is the consistent observation that animals lack precision in adjusting trajectories toward the submerged escape platform in

---

[b] Some local views need to be stored for subsequent place identification while navigating in the environment (see section 3).

the Morris task. However, as I have emphasized in Chapter 7, we do not know whether this inaccuracy is due to an imprecise encoding of the goal location during acquisition or to a sensorimotor impairment in steering the trajectory to the goal location. In this respect, the study of exploratory strategies in parietal animals may help toward understanding at what stage(s) of spatial processing the associative parietal cortex plays a prominent role.

The level which raises the most ticklish questions is that of the representations. If the role of the hippocampal formation in spatial representations is indisputable, the precise nature of what the hippocampal system does and *how long* it ensures the storage of spatial memories are the subjects of lively debates. Indeed, the properties of place cells and head direction cells, together with their demonstrated sensitivity to environmental and movement-generated inputs, allows us to postulate that most of the initial, genuinely spatial processing is performed by the hippocampal formation. Whereas some authors (e.g., O'Keefe and Nadel, 1978; O'Keefe, 1991a and b) minimize the importance of intermediate levels of spatial processing, between perception and representation, others (e.g., McNaughton et al., 1989a; Poucet, 1993) assign an important spatial function to the parietal cortex. This concurs with my opinion in view of the data that have resulted from lesion and electrophysiological studies.

Various models concerning the role of the hippocampal system – whether it acts in cooperation with other brain regions or alone – have been presented in the preceding chapter. Among these proposals, the general model of memory put forward by McClelland et al. (1995) accounts for many data and is of special relevance for generating further hypotheses about spatial processing. Not only does it account for progressive retrograde amnesia following hippocampal damage, but, more importantly for our purposes, it proposes an explanation of how specific representations elaborated by the hippocampal system could contribute to the formation of more abstract "structures". Applied to spatial representations, McClelland et al.'s model provides a hypothesis about how, by successive experiences of spatial situations interleaved with other types of experience, abstract representations and rules for spatial processing could be progressively elaborated in the neocortex. In the case when a context-specific situation is repeatedly experienced, then, "... this information, too, would gradually become incorporated into the connection weights in the neocortical system and would sustain performance in an episodic memory task". (McClelland et al., 1995). Thus, in this theory, whether they are episodic (context-specific) or abstract, memories become stored in the long term in the neocortical system.

However, as emphasized by Nadel (1991; see Chapter 9), the fact that spatial performance can be spared when hippocampal lesions are made after a sufficiently long post-training delay does not imply that the hippocampus has no role to play after that delay. It merely points to the fact that neocortical areas can ensure the achievement of effective performance. Therefore, one cannot rule out the possibility that context-specific spatial memories are also permanently stored in the hippocampus. One way to test such an hypothesis would be to examine the effects on spatial performance of neocortical lesions (in parietal and prefrontal areas) made after a long post-training delay.

7. Concluding Remarks

To summarize in a few words the major conclusions that can be drawn from the host of data and theories discussed in this book, I would like to state the following points.

First, animals' spatial representations do have the properties that Tolman, and later O'Keefe and Nadel (1978), have assigned to cognitive maps. The most important attributed property is *behavioral flexibility*, because of its adaptive value to the organism. However, this statement concerns only spatially restricted situations such as those used in laboratories. It is not clear, at the present time, whether complex spatial computations can be performed by animals in situations which encompass visually disconnected regions of space. Further experiments in the same vein as those conducted by Sutherland et al. (1987) and Schenk and Gafner (in preparation) are required in order to definitely rule out the possibility that simple mechanisms have been implemented and to generalize the statement that animals do have and use cognitive maps whatever the size and complexity of the environment in which they are moving around.

Second, I would argue that one of the most important functions of cognitive maps is to constitute a spatial integrating system involved in the organization of how new incoming information is gathered. Such an organization can be analyzed: exploratory activity is an observable phenomenon. Therefore, the study of its dynamics appears likely to provide insight into the representations that control information gathering. This approach would also allow precise determination of the function of various brain structures, in particular those at the interface of action (including perceptual activity) and representation, during the acquisition and use of spatial information.

Obviously, the spatial integrating system (abstract representations and general rules for processing space) may be at an embryonic stage of development in rodents by comparison with human and non-human primates, corresponding to the respective development of the neocortical mantle. However, data generated by behavioral studies strongly suggest that general rules are implemented by rodents for the processing of space. Indeed, an abstract form of processing occurs as soon as generalization takes place.

Finally, another point deserves to be underlined. As human subjects, we are naturally disposed to conceive spatial representations as survey maps of the environment. This tendency is certainly due to the fact that, very early in life, we have many opportunities (while learning geography at school, for instance) to examine symbolic survey maps. Perhaps this is not the form of the maps in animals' heads (except perhaps for flying animals which have access to a bird's eye view of the spatial layout of their surroundings). In this respect, the latin root ("mappa") of the word "map" is highly instructive, since it refers to a sheet, i.e., to a continuous covering extent (a sheet presumably without holes!). Thus, the adaptive and flexible properties of spatial representations, that allow the inferring of new pathways from an incomplete knowledge of the environment would come to fill the lacunar regions of the map according to the etymological meaning of the word.

REFERENCES

Aadland, J., Beatty, W.W., & Maki, R.H. (1985). Spatial memory of children and adults assessed in the radial maze. *Developmental Psychobiology, 18,* 163-172.

Akert, K. (1964). Comparative anatomy of the frontal cortex and thalamo-cortical connections. In J.M. Warren & K. Akert (Eds.), *The frontal granular cortex and behavior* (pp. 372-396). McGraw-Hill Book Company.

Ammassari-Teule, M., & Caprioli, A. (1985). Spatial learning and memory, maze running strategies and cholinergic mechanisms in two inbred strains of mice. *Behavioral Brain Research, 17,* 9-16.

Ammassari-Teule, M., & Durup, H. (1982). Spatial learning in golden hamsters: relationship between food-searching strategies and difficulty of the task. *Behavioural Processes, 7,* 352-365.

Ammassari-Teule, M., Hoffmann, H.J., & Rossi-Arnaud, C. (1992). Learning in inbred mice: strain specific abilities across three radial maze problems. *Behavioral Genetics, 23,* 405-412.

Ammassari-Teule, M., Tozzi, A., Rossi-Arnaud, C., Save, E., & Thinus-Blanc, C. (1995). Reactions to spatial and nonspatial change in two inbred strains of mice: Further evidence supporting the hippocampal dysfunction hypothesis in the DBA/2 strain. *Psychobiology, 23,* 284-289.

Amsel, A., & Rashotte, M.E. (1984). *Mechanisms of adaptive behavior. Clark L. Hull's theoretical papers, with commentary*. New York: Columbia University Press.

Andersen, P., Bliss, T.V., & Skrede, K. (1971). Lamellar organization of the hippocampal excitatory pathways. *Experimental Brain Research, 13,* 222-238.

Andersen, R.A.G., Essick, K., & Siegel, R.M. (1985). Encoding of spatial location by posterior parietal neurons. *Science, 230,* 456-458.

Appleyard, D. (1969). Why buildings are known? *Environment and Behavior, 1,* 131-156.

Appleyard, D. (1970). Styles and methods of structuring a city. *Environment and Behavior, 2,* 100-118.

Arolfo, M.P., Nerad, L., Schenk, F., & Bures, J. (1994). Absence of snapshot memory of the target view interferes with place navigation learning by rats in the water maze. *Behavioral Neuroscience, 108,* 308-316.

Baddeley, A.D. (1986). *Working memory*. Oxford: Oxford University Press.

Baddeley, A.D. (1993). *La mémoire humaine. Théorie et pratique*. Grenoble: Presses Universitaires de Grenoble.

Baker, R.R. (1987). Accuracy of map-building and navigation by humans during "natural" exploration: relative roles of magnetoreception and vision. In P.

Ellen & C. Thinus-Blanc (Eds.), *Cognitive processes and spatial orientation in animal and man* (Vol. 1, pp. 217-232). Dordrecht, Neth.: Martinus-Nijhoff.

Balda, R.P. (1987). Avian impacts on pinyon-juniper woodlands. In R. Everett (Ed.), *Proceedings Pinyon-Juniper Conference* (pp. 525-533). U.S. Forest Serv. Gen. Techn. Rep., INT-215.

Barea, E.J., Hickey, M.R., & Smith, D.C. (1988). Hippocampal inactivation effects on spatial memory. *Society for Neuroscience Abstracts, 14,* 233.

Barnes, C.A. (1979). Memory deficits associated with senescence: a neurophysiological and behavioral study in the rat. *Journal of Comparative and Physiological Psychology, 93,* 74-104.

Barnes, C.A. (1988). Spatial learning and memory processes: the search for their neurobiological mechanisms in the rat. *Trends in Neuroscience, 11,* 163-169.

Batuev, A.S., Kursina, N.P., & Shutov, A.P. (1990). Unit activity of the medial wall of the frontal cortex during delayed performance in rats. *Behavioural Brain Research, 41,* 95-102.

Becker, J.T., Walker, J.A., & Olton, D.S. (1980). Neuroanatomical bases of spatial memory. *Brain Research, 200,* 307-321.

Benhamou, S. (1989). An olfactory orientation model for mammal's movements in their home range. *Journal of Theoretical Biology, 139,* 379-388.

Benhamou, S., & Poucet, B. (1996). A comparative analysis of spatial memory processes. *Behavioural Processes, 35*, 113-126.

Bennett, T.L. (1970). Hippocampal EEG correlates of behaviour. *Electroencephalography and Clinical Neurophysiology, 28,* 17-23.

Berlyne, D.E. (1950). Novelty and curiosity as determinants of exploratory behaviour. *British Journal of Psychology, 41,* 68-80.

Berlyne, D.E. (1960). *Conflict, arousal, and curiosity.* New York: McGraw-Hill.

Berlyne, D.E. (1966). Curiosity and exploration. *Science, 153,* 25-33.

Best, P.J., & Thompson, L.T. (1989). Persistence, reticence, and opportunism of place-field activity in hippocampal neurons. *Psychobiology, 17,* 236-246.

Blancheteau, M. (1969). *L'orientation spatiale chez l'animal, ses indices et ses repères.* Paris: Editions du CNRS.

Blancheteau, M., & Le Lorec, A. (1972). Raccourci et détour chez le Rat: durée, vitesse et longueur des parcours. *L'Année Psychologique, 72,* 7-16.

Blodgett, H.C. (1929). The effect of the introduction of reward upon the maze performance of rats. *University of California Publications in Psychology, 4,* 120.

Bolles, R.C. (1975). *Learning theory.* New York: Holt, Rinehart and Winston.

Boyd, M.G., & Thomas, R.K. (1977). Posterior association cortex lesion in rats: mazes, patterns discrimination, and reversal learning. *Physiological Psychology, 4,* 455-461.

Brito, G.N.O., & Brito, L.S.O. (1990). Septohippocampal system and the prelimbic sector of frontal cortex: a neurophysiological battery analysis in the rat. *Behavioral Brain Research, 36,* 127-146.

Brito, G.N.O., Thomas, G.J., Davis, B.J., & Gingold, S.I. (1982). Prelimbic cortex, mediodorsal thalamus, septum and delayed alternation in rats. *Experimental Brain Research, 46,* 52-58.

Brown, M.F. (1992). Does a cognitive map guide choices in the radial-arm maze? *Journal of Experimental Psychology: Animal Behavior Processes, 18,* 56-66.

Brown, M.F., & Cook, R.G. (1986). Within-trial dynamics of radial arm maze performance in rats. *Learning and Motivation, 17,* 190-205.

Brown, M.F., Wheeler, A.E., & Riley, D.A. (1989). Evidence for a shift in the choice criterion of rats in a 12-arm radial maze. *Animal Learning and Behavior, 17,* 12-20.

Brown, V.J., Bowman, E.M., & Robbins, T.W. (1991). Response-related deficits following unilateral lesions of the medial agranular cortex of the rat. *Behavioral Neuroscience, 105,* 567-578.

Buhot, M.-C., & Poucet, H. (1987). Role of the spatial structure in multiple choice problem-solving by golden hamsters. In P. Ellen & C. Thinus-Blanc (Eds.), *Cognitive processes and spatial orientation in animal and man* (pp. 124-134. Dordrecht, Neth.: Martinus-Nijhoff.

Buhot, M.-C., & Teule, M. (1971). Apprentissage d'élimination progressive chez le Hamster doré (*Mesocricetus auratus*). *Cahiers de Psychologie, 14,* 135-165.

Buresova, O. (1980). Spatial memory and instrumental conditioning. *Acta Neurobiologia Experientia, 40,* 51-66.

Burgess, N., O'Keefe, J., & Recce, M. (1993). Using hippocampal "place cells" for navigation, exploiting phase coding. In S. Hanson, J. Cowan, & L. Giles (Eds.), *Advances in neural information processing systems 5* (pp. 929-936). San Mateo, CA: Morgan Kaufmann.

Carey, D.P., Booth, L.A., & Goodale, M.A. (1987). Can gerbils with lesions of primary visual cortex use retinal image size to estimate distance in a jumping task? *Society for Neuroscience Abstracts, 13,* 629.

Carmen, J.B., Cowan, W.M., & Powell, T.P.S. (1964). The cortical projection upon the claustrum. *Journal of Neurological and Neurosurgical Psychiatry, 27,* 46-51.

Cartwright, B.A., & Collet, T.S. (1982). How honey bees use landmarks to guide their return to a food source. *Nature, 295,* 560-564.

Cartwright, B.A., & Collett, T.S. (1983). Landmark learning in bees: experiments and models. *Journal of Comparative Physiology, 151,* 521-543.

Chapuis, N. (1987). Detour and shortcut abilities in several species of mammals. In P. Ellen & C. Thinus-Blanc (Eds.), *Cognitive processes and spatial orientation in animal and man* (Vol. 1, pp. 97-106). Dordrecht, Netherl.: Martinus Nijhoff.

Chapuis, N. (1988). *Les opérations structurantes dans la connaissance de l'espace chez les Mammifères: détour, raccourci et retour.* Doctoral Dissertation, Université d'Aix-Marseille II.

Chapuis, N., & Varlet, C. (1987). Shortcuts by dogs in natural surroundings. *Quarterly Journal of Experimental Psychology, 39B,* 49-64.

Chapuis, N., Durup, M., & Thinus-Blanc, C. (1987). The role of exploratory experience in a shorcut task by golden hamsters. *Animal Learning and Behavior, 15,* 174-178.

Chapuis, N., Thinus-Blanc, C., & Poucet, B. (1983). Dissociation of mechanisms involved in dog's oriented displacements. *Quarterly Journal of Experimental Psychology, 35B,* 213-219.

Cheal, M.-L. (1978). Stimulus-elicited investigation in the Mongolian gerbil (*Meriones unguiculatus*). *The Journal of Biological Psychology, 20,* 26-32.

Chen, L.L., Lin, L.-H., Barnes, C.A., & McNaughton, B.L. (1994a). Head-direction cells in the rat posterior cortex. II. Contributions of visual and ideothetic information to the directional firing. *Experimental Brain Research, 101,* 24-34.

Chen, L.L., Lin, L.-H., Green, E.J., Barnes, C.A., & McNaughton, B.L. (1994b). Head-direction cells in the rat posterior cortex. I. Anatomical distribution and behavioral modulation. *Experimental Brain Research, 101,* 8-23.

Chen, L.L., & McNaughton, B.L. (1988). Spatially selective discharge of vision and movement modulated posterior parietal neurons in the rat. *Society for Neuroscience Abstracts, 14,* 818.

Cheng, K. (1986). A purely geometric module in the rat's spatial representation. *Cognition, 23,* 149-178.

Cheng, K. (1987). Rats use the geometry of surfaces for navigation. In P. Ellen & C. Thinus-Blanc (Eds.), *Cognitive processes and spatial orientation in animal and man* (Vol. 1, pp. 153-159). Dordrecht, Neth.: Martinus-Nijhoff.

Cheng, K., & Gallistel, C.R. (1984). Testing the geometric power of an animal's spatial representation. In H.L. Roitblat, T.G. Bever, & H.S. Terrace (Eds.), *Animal cognition* (pp. 409-423). Hillsdale, NJ: Lawrence Erlbaum Associates.

Collett, T.S., Cartwright, B.A., & Smith, B.A. (1986). Landmark learning and visuo-spatial memories in gerbils. *Journal of Comparative Physiology, 158,* 835-851.

Condé, F., Audinat, E., Maire-Lepoivre, E., & Crépel, F. (1990). Afferent connections of the medial frontal cortex of the rat. A study using retrograde transport of fluorescent dyes. I. Thalamic afferents. *Brain Research Bulletin, 24,* 341-354.

Cook, R.G., Brown, M.F., & Riley, D.A. (1985). Flexible memory processing by rats: use of prospective and retrospective information in the radial maze. *Journal of Experimental Psychology: Animal Behavior Processes, 11,* 453-469.

Corman, C.D., & Shafer, J.N. (1968). Open-field activity and exploration. *Psychonomic Science, 13,* 55-66.

Cotman, C.W., & Lynch, G.S. (1989). The neurobiology of learning and memory. *Cognition, 33,* 201-241.

Cowan, P.E. (1976). The new object reaction of Rattus rattus L.: the relative importance of various cues. *Behavioral Biology, 16,* 31-44.

Cowie, R.J., Krebs, J.R., & Sherry, D.F. (1981). Food-storing by marsh tits. *Animal Behaviour, 29,* 1252-1259.

Crannell, C.W. (1942). The choice point behavior of rats in a multiple path elimination problem. *Journal of Psychology, 13,* 201-222.

Cressant, A., Poucet, B., & Muller, R.U. (1996). The spatial arrangement of 3-dimensional objects inside the arena influences their control over positional firing of place cells. *Society for Neuroscience Abstracts* (in press).

Crusio, W.E., Schwegler, H., & Lipp, H.P. (1987). Radial maze performance and structural variation of the hippocampus in mice: a correlation with mossy fiber distribution. *Brain Research, 425,* 182-185.

Dale, R.H.I. (1980). The role of vision in the rat's radial maze performance. *Dissertation Abstracts International, 40,* 5047B.

Dale, R.H.I., & Innis, N. (1980). *Spatial memory without vision: radial maze performance of blind rats*. University of Western Ontario Research Bulletin, n° 523.

Dale, R.H.I., & Roberts, W.A. (1986). Variations in radial maze performance under different levels of food and water deprivation. *Animal Learning and Behavior, 14,* 60-64.

Dallal, N.L., & Meck, W.H. (1990). Hierarchical structures: chunking by food type facilitates spatial memory. *Journal of Experimental Psychology: Animal Behavior Processes, 16,* 69-84.

Damasio, A.R. (1985). Disorders of complex visual processing. In M.M. Mesulam (Ed.), *Principles of behavioral neurology*. Philadelphia, PA: Davis.

Damasio, A.R. (1989a). The brain binds entities and events by multiregional activation from convergence zones. *Neural Computation, 1,* 123-132.

Damasio, A.R. (1989b). Time-locked multiregional retroactivation: a system-level proposal for the neural substrates of recall and recognition. *Cognition, 33,* 25-62.

Damasio, A.R. (1990). Synchronous activation in multiple cortical regions: a mechanism for recall. *Seminars in Neurosciences, 2,* 287-296.

Davis, B.K., & McDaniel, W.F. (1993). Visual memory and visual spatial functions in the rat following parietal and temporal cortex injuries. *Physiology and Behavior, 53,* 145-151.

De Renzi, E. (1982). *Disorders of space exploration and cognition.* Chichester: John Wiley and Sons.

Dean, P. (1990). Sensory cortex: visual perceptual functions. In B. Kolb & R.C. Tees (Eds.), *The cerebral cortex of the rat* (pp. 275-307). Cambridge, MA: The MIT Press.

Dean, P., & Redgrave, P. (1984). The superior colliculus and visual neglect in rat and hamster. I. Behavioural evidence. II. Possible mechanisms. III. Functional implications. *Brain Research Review, 8,* 129-141, 143-153, 155-163.

Dember, W.N. (1956). Response by the rat to environmental change. *Journal of Comparative and Physiological Psychology, 49,* 93-95.

Denis, M. (1989). *Image et cognition.* Paris: P.U.F.

Devan, B.D., Blank, G.S., & Petri, H.L. (1992). Place navigation in the Morris water task: effects of reduced platform interval lighting and pseudorandom platform positioning. *Psychobiology, 20,* 120-126.

Dickinson, A. (1980). *Contemporary animal learning theory.* Cambridge, UK: Cambridge University Press.

Diez-Chamizo, V., Sterio, D., & Mackintosh, N.J. (1985). Blocking and overshadowing between intra-maze and extra-maze cues: a test of the independence of locale and guidance system. *Quarterly Journal of Experimental Psychology, 37B,* 235-253.

Dimattia, B.V., & Kesner, R.P. (1988a). Role of the posterior parietal association cortex in the processing of spatial event information. *Behavioral Neuroscience, 102,* 397-403.

Dimattia, B.V., & Kesner, R.P. (1988b). Spatial cognitive maps: differential role of the parietal cortex and hippocampal formation. *Behavioral Neuroscience, 102,* 471-480.

Divac, I., Kosmal, A., Björklund, A., & Lindvall, O. (1978). Subcortical projections to the prefrontal cortex in the rat as revealed by the horseradish peroxidase technique. *Neuroscience, 3,* 785-796.

Divac, I., Wilmark, R.G.E., & Gade, A. (1975). Spontaneous alternation in rats with lesions in the frontal lobes. An extension of the frontal lobe syndrome. *Physiological Psychology, 3,* 39-42.

Douglas, R.J. (1975). The development of hippocampal function: implications for theory and therapy. In R.L. Isaacson & K.H. Pribram (Eds.), *The hippocampus* (Vol. 2, pp. 309-326). New York: Academic Press.

Downs, R.M. (1981). Maps and mapping as metaphors for spatial representations. In L.S. Lyben, A.H. Patterson, & N. Newcombe (Eds.), *Spatial representations and behavior across the life span. Theory and applications* (pp. 143-166). London: Academic Press.

Eichenbaum, H., Otto, T., & Cohen, N. (1994). Two functional components of the hippocampal memory system. *Behavioral and Brain Sciences, 17,* 449-518.

Ellen, P. (1987). Cognitive mechanisms in animal problem-solving. In P. Ellen & C. Thinus-Blanc (Eds.), *Cognitive processes and spatial orientation in animal and man* (Vol. I, pp. 20-35). Dordrecht, Neth.: Martinus Nijhoff.

Ellen, P., Parko, E.M., Wages, C., Doherty, D., & Herrmann, T. (1982). Spatial problem solving by rats: exploration and cognitive maps. *Learning and Motivation, 13,* 81-94.

Ellen, P., Soteres, B.J., & Wages, C. (1984). Problem solving in the rat: piecemeal acquisition of cognitive maps. *Animal Learning and Behavior, 12,* 232-237.

Etienne, A. (1987). The control of short-distance homing in the golden hamster. In P. Ellen & C. Thinus-Blanc (Eds.), *Cognitive processes and spatial orientation in animal and man* (Vol. I, pp. 233-251). Dordrecht, Neth.: Martinus-Nijhoff.

Etienne, A., Joris-Lambert, S., Reverdin, B., & Teroni, E. (1993). Learning to recalibrate the role of dead reckoning and visual cues in spatial navigation. *Annual Learning and Behavior, 21,* 266-280.

Etienne, A.E., Maurer, R., & Saucy, F. (1988). Limitations in the assessment of path dependent information. *Behaviour, 106,* 81-111.

Etienne, A.S., Joris, S., Maurer, R., & Teroni, E. (1990a). Enhancing the impact of visual extra-maze cues in a spatial orientation task. *Behavioural Brain Research, 38,* 199-210.

Etienne, A.S., Teroni, E., Hurni, C., & Portenier, V. (1990b). The effect of a single light cue on homing behaviour of the golden hamster. *Animal Behaviour, 39,* 17-41.

Fabrigoule, C. (1974). Recherches expérimentales sur l'apprentissage spatial chez le chien. Apprentissage d'élimination progressive. *Cahiers de Psychologie, 17,* 91-110.

Fabrigoule, C. (1987). Etude des processus cognitifs impliqués dans l'apprentissage spatial chez le chien et le renard. Doctoral Dissertation, Université d'Aix-Marseille II.

Farah, M.J. (1984). The neurological basis of mental imagery: a componential analysis. *Cognition, 18,* 245-272.

Farah, M.J. (1989). The neuropsychology of mental imagery. In A. Damasio (Ed.), *Handbook of neuropsychology: disorders of visual processing* (Vol. II, pp. 395-413). Amsterdam: Elsevier.

Farah, M.J., Peronnet, F., Gonon, M.-A., & Giard, M.-H. (1988). Electrophysiological evidence for a shared representational medium for visual images and visual percepts. *Journal of Experimental Psychology: General, 117,* 248-257.

Ferrier, R.J., & Cooper, R.M. (1976). Striate cortex ablation and spatial vision. *Brain Research, 106,* 71-85.

Finke, R.A. (1980). Levels of equivalence in imagery and perception. *Psychological Review, 87,* 113-132.

Finke, R.A. (1985). Theories relating mental imagery to perception. *Psychological Bulletin, 87,* 236-259.

Foerster, R. (1890). Ueber Rindenblindheit. Albrecht v. Graefe's. *Archives of Ophtalmology, 36,* 94-108.

Foreman, N. (1983). Distractibility following simultaneous bilateral lesions of the superior colliculus or medial frontal cortex in the rat. *Behavioural Brain Research, 8,* 177-194.

Foreman, N. (1985). Algorithmic responding on the radial maze in rats does not always imply the absence of spatial encoding. *Quarterly Journal of Experimental Psychology, 37B,* 333-358.

Foreman, N. (1996). Twenty years of the radial arm maze. In N. Foreman & R. Gillett (Eds.), *A handbook of spatial research paradigms and methodologies*. Hove, UK: Lawrence Erlbaum Associates (in press).

Foreman, N., Arber, M., & Savage, J. (1984). Spatial memory in preschool infants. *Developmental Psychobiology, 17,* 129-137.

Foreman, N., Save, E., Thinus-Blanc, C., & Buhot, M.-C. (1992). Visually guided locomotion, distractibility, and the missing stimulus effect in hooded rats with unilateral or bilateral lesions of parietal cortex. *Behavioral Neuroscience, 106,* 529-538.

Foreman, N., & Stevens, R. (1982). Visual lesions and radial maze performance in rats. *Behavioural and Neural Biology, 36,* 126-136.

Foreman, N., & Stevens, R. (1987). Relationships between the superior colliculus and hippocampus: neural and behavioral considerations. *Behavioral and Brain Sciences, 10,* 101-152.

Foreman, N., & Thinus-Blanc, C. (1987). The "missing stimulus" effect in hooded rats: gross asymetry in the Sokolovian orienting response. *Psychobiology, 15,* 265-271.

Foster, T.C., Castro, C.A., & McNaughton, B.L. (1989). Spatial selectivity of rat hippocampal neurons is dependent on preparedness for movement. *Science, 244,* 1580-1582.

Fregnac, Y., Shultz, D., Thorpe, S., & Bienenstock, E. (1988). A cellular analogue of visual cortical plasticity. *Nature, 333,* 367-370.

Füller, E., Kowalski, U., & Wiltshko, R. (1983). Orientation of homing pigeons: compass orientation vs. piloting by familiar landmarks. *Journal of Comparative Physiology, 153,* 55-58.

Gallagher, M., Burwell, R., & Burchinal, M. (1993). Severity of spatial learning impairment in aging: development of a learning index for performance in the Morris water maze. *Behavioral Neuroscience, 107,* 618-626.

Gallistel, C.R. (1990). *The organization of learning*. Cambridge, MA: Bradford Books/MIT Press.

Gaunet, F., Martinez, J.L., & Thinus-Blanc, C. (1996). Early blind subject's spatial abilities in the manipulatory space: a study of exploratory strategies and reaction to change performance. *Perception* (in press).

Gaunet, F., & Thinus-Blanc, C. (1996). Early blind subject's spatial abilities in the locomotor space: a study of exploratory strategies and reaction to change performance. *Perception* (in press).

Geier, F.M., Levine, M., & Tolman, E.C. (1941). Individual differences in emotionality, hypothesis formation, vicarious trial and error, and visual discrimination learning in rats. *Comparative Psychology Monograph, 17.*

Georgakopoulos, J., & Etienne, A. (1994). Identifying location by dead reckoning and external cues. *Behavioural Processes, 31,* 57-74.

Gibson, J.J. (1979). *The ecological approach to visual perception.* Boston, MA: Houghton-Mifflin.

Glanzer, M. (1953). Stimulus satiation: an explanation of spontaneous alternation and related phenomena. *Psychological Review, 60,* 257-268.

Goddard, G.V. (1980). Component properties of the memory machine. In *The nature of thought: essays in honour of D.O. Hebb.* Hillsdale, NJ: Lawrence Erlbaum Associates.

Goldenberg, G., Podreka, I., Steiner, M., Suess, E., Deecke, L., & Willmes, K. (1988). Pattern of regional cerebral blood flow related to visual and motor imagery: results of emission computerized tomography. In M. Denis, J. Engelkamp, & J.T.E. Richardson (Eds.), *Cognitive and neuropsychological approaches to mental imagery* (pp. 363-373). Dordrecht: Martinus-Nijhoff.

Goldenberg, G., Podreka, J., Steiner, M., & Willmes, K. (1987). Patterns of regional cerebral blood flow related to memorizing of high and low imagery worlds. An emission computer tomography study. *Neuropsychologia, 25,* 473-485.

Goldstein, L.H., & Oakley, D.A. (1987). Visual discrimination in the absence of visual cortex. *Behavioural Brain Research, 24,* 181-193.

Goodale, M.A. (1983). Neural mechanisms of visual orientation in rodents: targets versus places. In A. Hein & M. Jeannerod (Eds.), *Spatially oriented behavior* (pp. 35-61). New York: Springer-Verlag.

Goodale, M.A., & Carey, D.P. (1990). The role of the cerebral cortex in visuomotor control. In B. Kolb & R.C. Tees (Eds.), *The cerebral cortex of the rat* (pp. 309-340). Cambridge, MA: The MIT Press.

Goodale, M.A., & Dale, R.H.I. (1981). Radial-maze performance in the rat following lesions of posterior neocortex. *Behavioural Brain Research, 3,* 273-288.

Goodale, M.A., & Milner, A.D. (1982). Fractionating orientation behavior in the rodent. In D. Ingle, M. Goodale, & R. Mansfield (Eds.), *Analysis of visual behavior* (pp. 267-299). Cambridge, MA: The MIT Press.

Goodale, M.A., & Murison, R.C.C. (1975). The effects of lesions of the superior colliculus on locomotor orientation and the orienting reflex in the rat. *Brain Research, 88,* 243-25.

Goodridge, J.P., & Taube, J.S. (1995). Preferential use of the landmark navigational system by head direction cells in rats. *Behavioral Neuroscience, 109,* 49-61.

Granon, S., & Poucet, B. (1995). Medial prefrontal lesions in the rat and spatial navigation. Evidence for impaired planning. *Behavioral Neuroscience, 109* (3), 474-484.

Granon, S., Save, E., Buhot, M.C., & Poucet, B. (1996). Effortful information processing in a spontaneous spatial situation by rats with medial prefrontal lesions. *Behavioural Brain Research* (in press).

Granon, S., Vidal, C., Thinus-Blanc, C., Changeux, J.P., & Poucet, B. (1994). Working memory, response selection, and effortful processing in rats with medial prefrontal lesions. *Behavioral Neuroscience, 108* (5), 883-891.

Gray, J.A. (1982). *The neuropsychology of anxiety: an enquiry into the functions of the septo-hippocampal system.* Oxford: Oxford University Press.

Gray, J.A., & McNaughton, N. (1983). Comparison between the behavioural effects of septal and hippocampal lesions: a review. *Neuroscience and Biobehavioral Reviews, 7,* 119-188.

Green, E.J., McNaughton, B.L., & Barnes, C.A. (1990). Exploration-dependent modulation of evoked responses in Fascia dentata: dissociation of motor, EEG, and sensory factors and evidence for a synaptic efficacy change. *The Journal of Neuroscience, 10,* 1455-1471.

Grobéty, M.-C., & Schenk, F. (1992). The influence of spatial irregularity upon radial-maze performance in the rat. *Animal Learning and Behavior, 20,* 393-400.

Groenewegen, H.J. (1988). Organization of the afferent connections of the mediodorsal thalamic nucleus in the rat, related to the mediodorsal prefrontal topography. *Neuroscience, 24,* 379-431.

Groenewegen, H.J., & Berender, H.W. (1994). Anatomical realtionships between the prefrontal cortex and the basal ganglia in the rat. In A.M. Thierry, J. Glowinsky, P.S. Goldman-Rakic, & Y. Christen (Eds.), *Motor and cognitive functions of the prefrontal cortex* (pp. 51-77). Berlin, Heidelberg: Fondation Ipsen, Springer Verlag.

Guldin, W., Mirring, S., & Grüsser, O.-J. (1992). Monosynaptic input from the cerebral cortex to the vestibular brainstem nuclei in the rat. *Society for Neuroscience Abstracts, 18,* 510.

Guthrie, E.R. (1952). *The psychology of learning*. New York: Harper and Row.

Haig, K.A., Rawlins, J.N.P., Olton, D.S., Mead, A., & Taylor, B. (1983). Food searching strategies of rats: variables affecting the relative strength of stay and shift strategies. *Journal of Experimental Psychology: Animal Behavior Processes, 9,* 337-348.

Hall, C.S. (1934). Emotional behavior in the rat. I: Defecation and urina as measures of individual differences in emotionality. *Journal of Comparative Psychology, 18,* 385-403.

Herrmann, T., Poucet, B., & Ellen, P. (1985a). Spatial problem solving in a dual runway task by normal and septal rats. *Behavioral Neuroscience, 99,* 631-637.

Herrmann, T., Poucet, B., & Ellen, P. (1985b).Spatial problem solving in the rat following medial frontal lesions. *Physiological Psychology, 13,* 21-25.

Hill, A.J. (1978). First occurrence of hippocampal spatial firing in a new environment. *Experimental Neurology, 62,* 282-297.

Hilton, S.C., & Krebs, J.R. (1991). Spatial memory of four species of Parus: performance in an open-field analogue of a radial maze. *Quarterly Journal of Experimental Psychology, 42B,* 345-368

Hjorth-Simonsen, A. (1972). Projection of the lateral part of the entorhinal area to the hippocampus and fascia dentata. *Journal of Comparative Neurology, 146,* 219-232.

Hjorth-Simonsen, A., & Jeune, B. (1972). Origin and termination of the hippocampal perforant path in the rat studied by silver impregnation. *Journal of Comparative Neurology, 144,* 215-232.

Honig, W.K. (1978). Studies of working memory in the pigeon. In S.H. Hulse & W.K. Honig (Eds.), *Cognitive processes in animal behavior*. Hillsdale, NJ: Lawrence Erlbaum Associates.

Hudson, B. (1948). *One trial learning: a study of the avoidance behavior of the rat.* Doctoral Dissertation, University of California, Berkeley, CA.

Hull, C.L. (1930). Knowledge and purpose as habit mechanisms. *Psychological Review, 37,* 511-525.

Hull, C.L. (1932). The goal gradient hypothesis and maze learning. *Psychological Review, 39,* 25-43.

Hull, C.L. (1934a). The concept of the habit-family hierarchy and maze learning: Part 1. *Psychological Review, 41,* 33-54.

Hull, C.L. (1934b). The concept of the habit-family hierarchy and maze learning: Part 2. *Psychological Review, 41,* 33-54.

Hull, C.L. (1938). The goal-gradient hypothesis applied to some "field-force" problems in the behavior of young children. *Psychological Review, 45,* 271-299.

Hull, C.L. (1943). *Principles of behavior: an introduction to behavior theory.* New York: Appleton-Century-Crofts.

Hull, C.L. (1952). *A behavior system.* New Haven: Yale University Press.

Ilersich, T.J., Mazmanian, D.S., & Roberts, W.A. (1988). Foraging for covered and uncovered food on a radial maze. *Animal Learning and Behavior, 16,* 388-394.

Innis, N.K., & Macgillivray, M. (1987). Radial maze performance under food and water deprivation. *Behavioural Processes, 15,* 167-179.

Jackson, L.L. (1943). VTE on an elevated T-maze. *Journal of Comparative Psychology, 36,* 99-107.

Jarrard, L.E., Okaichi, H., Steward, O., & Goldschmidt, R.B. (1984). On the role of hippocampal connections in the performance of place and cue tasks: comparisons with damage to hippocampus. *Behavioral Neuroscience, 98,* 946-954.

Jay, T.M., & Witter, M.P. (1991). Distribution of hippocampal CA1 and subicular efferents in the prefrontal cortex of the rat studied by means of anterograde transport of Phaseolus vulgaris-leucoagglutinin. *Journal of Comparative Neurology, 313,* 574-586.

Jay, T.M., Glowinsky, J., & Thierry, A.-M. (1989). Selectivity of the hippocampal projection to the prelimbic area of the prefrontal cortex in the rat. *Brain Research, 505,* 337-340.

Jones, C., McGhee, R., & Wilkie, D.M. (1990). Hamsters (*Mesocricetus auratus*) use spatial memory in foraging for food to hoard. *Behavioural Processes, 21,* 179-187.

Joubert, A., & Vauclair, J. (1986). Reaction to novel objects in a troop af Guinea baboons: approach and manipulation. *Behaviour, 96,* 92-104.

Jung, M.W., & McNaughton, B.L. (1993). Spatial selectivity of unit activity in the hippocampal granule layer. *Hippocampus, 3,* 165-182.

Jung, M.W., Wiener, S.I., & McNaughton, B.L. (1994). Comparison of spatial firing characteristics of units in dorsal and ventral hippocampus of the rat. *The Journal of Neuroscience, 14,* 7347-7356.

Kant, I. (1963). *Critique of pure reason* (2nd edition, first published in German, 1787). London: MacMillan.

Kaplan, S. (1982). Lost in Chelm: maladaptive behavior in an adaptive model. *Behavioral and Brain Sciences, 5,* 643-644

Keith, J.R., & McVety, K.M. (1988). Latent place learning in a novel environment and the influences of prior training in rats. *Psychobiology, 16,* 146-151.

Kelsey, B.A., & Landry, J.E. (1988). Medial septal lesions disrupt spatial mapping ability in rats. *Behavioral Neuroscience, 102,* 289-293.

Kesner, R.P. (1980). An attribute analysis of memory: the role of the hippocampus. *Physiological Psychology, 8,* 189-197.

Kesner, R.P. (1984). The neurobiology of memory: implicit and explicit assumptions. In J.L. McGaugh, G. Lynch, & N.M. Weinberger (Eds.), *Neurobiology of learning and memory* (pp. 111-118). New York: Guilford Press.

Kesner, R.P. (1986). Neurobiological views of memory. In J.L. Martinez Jr and R.P. Kesner (Eds.), *Learning and memory: a biological view* (pp. 399-438). London: Academic Press Inc., Harcourt Brace Jovanovich Publ.

Kesner, R.P. (1989). Retrospective and prospective coding of information: role of the medial prefrontal cortex. *Experimental Brain Research, 74,* 163-167.

Kesner, R.P., DiMattia, B.V., & Crutcher, K.A. (1987). Evidence for neocortical involvement in reference memory. *Behavioral and Neural Biology, 47,* 40-53.

Kesner, R.P., Farnworth, G., & Dimattia, B.V. (1989). Double dissociation of egocentric and allocentric space following medial prefrontal and parietal cortex lesions in the rat. *Behavioral Neuroscience, 103,* 956-961.

Knierim, J.J., Kudrimoti, H.S., & McNaughton, B.L. (1995). Place cells, head direction cells, and the learning of landmark stability. *The Journal of Neuroscience, 15,* 1648-1659.

Köhler, W. (1927). *L'intelligence des Singes supérieurs.* Paris: Librairie Félix Alcan. (Translation and edition of 1964.)

Kolb, B. (1984). Functions of the frontal cortex of the rat: a comparative review. *Brain Research Reviews, 8,* 65-98.

Kolb, B. (1990a). Posterior parietal and temporal association cortex. In B. Kolb & R.C. Tees (Eds.), *The cerebral cortex of the rat* (pp. 459-471). Cambridge, MA: MIT Press.

Kolb, B (1990b). Prefrontal cortex. In B. Kolb & R.C. Tees (Eds.), *The cerebral cortex of the rat* (pp. 437-458). Cambridge, MA: MIT Press.

Kolb, B., Buhrman, K., & McDonald, R. (1989). Dissociation of prefrontal, posterior parietal, and temporal cortical regions to spatial navigation and recognition memory in the rat. *Society for Neuroscience Abstracts, 15,* 244.4.

Kolb, B., Pittman, K., Sutherland, R.J., & Whishaw, I.Q. (1982). Dissociation of the contributions of the prefrontal cortex of dorsomedial thalamic nucleus to spatially guided behavior in the rat. *Behavioural Brain Research, 6,* 365-378.

Kolb, B., Sutherland, R.J., & Whishaw, I.Q. (1983). A comparison of the contributions of the frontal and parietal association cortex to spatial localization in rats. *Behavioral Neuroscience, 97,* 13-27.

Kolb, B., & Walkey, J. (1987). Behavioural and anatomical studies of the posterior parietal cortex in the rat. *Behavioural Brain Research, 23,* 127-145.

Kolb, B., & Whishaw, I.Q. (1983). Dissociation of the contributions of the prefrontal, motor, and parietal cortex to the control of movement in the rat: an experimental review. *Canadian Journal of Psychology, 37,* 211-232.

Kosslyn, S.M. (1980). *Image and mind.* Cambridge, MA: Harvard University Press.

Kosslyn, S.M., & Schwartz, S.P. (1977). A simulation of visual imagery. *Cognitive Science, 1,* 265-295.

Krebs, J.R., & McCleery, R.H. (1984). Optimization in behavioural ecology. In J.R. Krebs & N.B. Davis (Eds.), *Behavioural ecology* (2nd edition, pp. 91-121). Sunderland, MA: Sinauer.

Krech, I. (1932). Hypotheses in rats. *Psychological Review, 39,* 516-532.

Krettek, J.K., & Price, J.L. (1977). The cortical projections of the mediodorsal nucleus and adjacent thalamic nuclei in the rat. *Journal of Comparative Neurology, 171,* 157-192.

Krieg, W.J.S. (1946). Connections of the cerebral cortex. I. The albino rat. A topography of the cortical areas. *Journal of Comparative Neurology, 84,* 221-275.

Krimm, M. (1994). *Rôle de l'expérience visuelle précoce et des aires visuelles corticales dans le traitement des informations spatiales.* Doctoral Dissertation, Université d'Aix-Marseille II.

Krushinskii, L.V. (1970). Objective study of elementary reasoning in animals. *Scientia, 64,* 1-13.

Kubie, J.L., Dayyani, R.U., Muller, B., Cohen, B., Major, E., & Sutherland, R.J. (1990). Hippocampal lesions in rats produce a temporally-graded retrograde amnesia on a spatial memory task. *Society for Neuroscience Abstracts, 16,* 1246.

Kubie, J.L., Dayyani, S., Sutherland, R.J., & Muller, R.U. (1989). Hippocampal lesions disrupt acquisition but not retention of navigational behavior in highly familiar environment. *Society for Neuroscience Abstracts, 15,* 609.

Kuipers, B. (1982). The "map in the head" metaphor. *Environment and Behavior, 14,* 202-220.

Kunka, M.G., Berstein, I.G., & Kubie, J.L. (1979). Short-term spatial memory in garter snakes. Paper presented at the Meeting of the Eastern Psychological Association, Philadelphia, April 1979.

Lachman, S.J. (1965). Behavior in a multiple-choice elimination problem involving five paths. *Journal of Psychology, 61,* 193-202.

Lachman, S.J. (1966). Stereotypy and variability of behavior in a complex learning situation. *Psychological Reports, 18,* 223-230.

Lachman, S.J. (1969a). Behavior in a complex learning situation involving five stimulus-differentiated paths. *Psychonomic Science, 17,* 36-37.

Lachman, S.J. (1969b). behavior in a three-path multiple choice elimination problem under conditions of overtraining. *Journal of Psychology, 73,* 101-109.

Lachman, S.J., & Brown, R. (1957). Behavior of a free choice multiple path elimination problem. *Journal of Psychology, 43,* 27-40.

Laroche, S., Jay, T.M., & Thierry, A.-M. (1990). Long-term potentiation in the prefrontal cortex following stimulation of the hippocampal CA1/ subicular region. *Neuroscience Letters, 114,* 184-190.

Larsen, J.K., & Divac, I. (1978). Selective ablations within the prefrontal cortex of the rat and performance of delayed alternation. *Physiological Psychology, 6,* 15-17.

Lashley, K.S. (1920). Studies of cerebral function in learning. *Psychobiology, 2,* 55-135.

Lashley, K. (1939). The mechanism of vision. XVI. The functioning of small remnants of the visual cortex. *Journal of Comparative Neurology, 70,* 45-67.

Lashley, K.S. (1943). Studies of cerebral function in learning. XII. Loss of the maze habit after occipital lesions in blind rats. *Journal of Comparative Neurology, 79,* 431-462.

Leonard, C.M. (1969). The prefrontal cortex of the rat. I. Cortical projection of the mediodorsal nucleus. II. Efferent connections. *Brain Research, 12,* 321-343.

Lewin, K. (1933). Vectors, cognitive processes, and M. Tolman's criticism. *Journal of General Psychology, 8,* 318-345.

Lewin, K. (1935). *A dynamic theory of personality: selected papers.* New York: McGraw-Hill.

Lieblich, I., & Arbib, M.A. (1982). Multiple representations of space underlying behavior. *Behavioral and Brain Sciences, 5,* 627-659.

Lockman, J.J. (1984). The development of detour ability during infancy. *Child Development, 55,* 482-491.

Lockman, J.J. (1987). Detour ability in infants and toddlers. In P. Ellen & C. Thinus-Blanc (Eds.), *Cognitive processes and spatial orientation in animal and man* (Vol. II, pp. 250-256). Dordrecht, Neth.: Martinus Nijhoff.

Lynch, K. (1960). *The image of the city.* Cambridge, MA: The MIT Press.

Maier, N.R.F. (1932). A study of orientation in the rat. *Journal of Comparative Psychology, 14,* 387-399.

Margules, J., & Gallistel, C.R. (1988). Heading in the rat: determination by environmental shape. *Animal Learning and Behavior, 16,* 404-410.

Matthews, D.B., White, A.M., Brush, E.D., & Best, P.J. (1995). Construction of the spatial cognitive map does not require active exploration of the environment. *Society for Neuroscience Abstracts, 21,* 2086.

Mazmanian, D.S., & Roberts, W.A. (1983). Spatial memory in rats under restricted viewing conditions. *Learning and Motivation, 14,* 123-139.

McClelland, J.L., McNaughton, B.L., & O'Reilly, R.C. (1995). Why there are complementary learning systems in the hippocampus and neocortex: insights from the successes and failures of connectionist modes of learning and memory. *Psychological Review, 102,* 419-457.

McDaniel, W.F., & Braucht, G.S. (1984). Pattern discrimination acquisition and performance following regionally limited posterior neocortical ablations. *IRCS Medical Science, 12,* 1086-1087.

McDaniel, W.F., & Skeel, R.L. (1993). Visual-spatial functions persit following temporal and posterior parietal cortex lesions in rat. *Physiology and Behavior, 53,* 199-203.

McDaniel, W.F., & Thomas, R.K. (1978). Temporal and parietal association cortex lesions and spatial and black-white reversal learning in the rat. *Physiological Psychology, 7,* 300-305.

McDaniel, W.F., & Wall, T.T. (1988). Visuospatial functions in the rat following injuries to striate, peristriate, and parietal neocortical sites. *Psychobiology, 16,* 251-260.

McDaniel, W.F., Via, J.D., Smith, J.S., Wells, D.L., FU, J.J., Bishop, J.F., Boyd, P.A., & Ledesma, H.M. (1995). Unilateral injury of posterior parietal cortex and spatial learning in hooded rats. *Behavioural Brain Research, 70,* 165-179.

McDaniel, W.F., Wildman, L.D., & Spears, R.H. (1979). Posterior association cortex and visual pattern discrimination in the rat. *Physiological Psychology, 7,* 241-244.

McNaughton, B.L. (1983). Activity dependent modulation of hippocampal synaptic efficacy: some implications for memory processes. In W. Seifert (Ed.), *Neurobiology of the hippocampus* (pp. 233-252). London: Academic Press.

McNaughton, B.L. (1987). Neural association of movement and space: preliminary steps toward a non-cartographic theory of spatial representation and learning. *Neuroscience Letters Suppl., 29,* S143-S144.

McNaughton, B.L. (1989a). Neural mechanisms for spatial computation and information storage. In L. Nadel, L.A. Cooper, P. Culicover, & R.M. Harmish (Eds.), *Neural connections and mental computations* (pp. 285-350). Cambridge, MA: MIT Press.

McNaughton, B.L. (1989b). The neurobiology of spatial computation and learning. In D.J. Stein (Ed.), *Lectures on complexity. Santa Fe Institute studies in the sciences of complexity* (pp. 389-437). Redwood, CA: Addison-Wesley.

McNaughton, B.L., Barnes, C.A., Meltzer, J., & Sutherland, R.J. (1989a). Hippocampal granule cells are necessary for normal spatial learning but not for spatially-selective pyramidal cell discharge. *Experimental Brain Research, 52,* 41-49.

McNaughton, B.L., Barnes, C.A., & O'Keefe, J. (1983). The contributions for position, direction, and velocity to single-unit activity in the hippocampus of freely-moving rats. *Experimental Brain Research, 52,* 41-49.

McNaughton, B.L., Barnes, C.A., Rao, G., Baldwin, J., & Rasmussen, M. (1986b). Long-term enhancement of hippocampal synaptic transmission and the acquisition of spatial information. *Journal of Neuroscience, 6,* 563-571.

McNaughton, B.L., Chen, L.L., & Markus, E.J. (1991). "Dead reckoning" landmark and the sense of direction: a neurophysiological and computational hypothesis. *Journal of Cognitive Neuroscience, 3,* 190-202.

McNaughton, B.L., Green, E., & Mizumori, S.J.Y. (1986a). Representation of body-motion trajectory by rat sensory-motor cortex neurons. *Society for Neuroscience Abstracts, 12,* 260.

McNaughton, B.L., Knierim, J.J., & Wilson, M.A. (1993a). Vector encoding and the vestibular foundations of spatial cognition: neurophysiological and computational mechanisms. In M.S. Gazzaniga (Ed.), *The cognitive neurosciences*. Cambridge, MA: The MIT Press.

McNaughton, B.L., Leonard, B., & Chen, L. (1989b). Cortical-hippocampal interactions and cognitive mapping: a hypothesis based on reintegration of the parietal and inferotemporal pathways for visual processing. *Psychobiology, 17,* 230-235.

McNaughton, B.L., Markus, E.J., Wilson, M.A., & Knierim, J.J. (1993b). Familiar landmarks can correct for cumulative error in the inertially based dead-reckoning system. *Society for Neuroscience Abstracts, 19,* 795.

McNaughton, B.L., & Morris, R.G.M. (1987). Hippocampal synaptic enhancement and information storage within a distributed memory system. *Trends in Neurosciences, 10,* 408-415.

McNaughton, B.L., & Nadel, L. (1990). Hebb-Marr networks and the neurobiological representation of action in space. In M.A. Gluck & D.E. Rumelhart (Eds.), *Neuroscience and connectionist theory* (pp. 1-64). Hillsdale, NJ: Lawrence Erlbaum Associates.

Menzel, E.W. (1978). Cognitive mapping in chimpanzees. In S.H. Hulse, H. Fowler, & W.K. Honig (Eds.), *Cognitive processes in animal behavior* (pp. 375-422). Hillsdale, NJ: Lawrence Erlbaum Associates.

Menzel, E.W., & Menzel, C.R. (1979). Cognitive developmental and social aspects of responsiveness to novel objects in a family group of marmosets (*Saguinus fuscicollis*). *Behaviour, 70,* 251-278.

Meyer, 0. (1900). Ein- und doppelseitige homonyme Hemianopsie mit Orientierungsstörungen. *Monatschriften Psychiatrische Neurologie, 8,* 440-456.

Miles, R., & Wong, R.K.S. (1986). Excitatory synaptic connections between CA3 neurons in the Guinea-pig hippocampus. *Journal of Physiology, 373,* 397-418.

Miles, R., & Wong, R.K.S. (1987). Latent synaptic pathways revealed after tetanic stimulation in the hippocampus. *Nature, 329,* 724-726.

Miller, M.W., & Vogt, B.A. (1984). Direct connections of rat visual cortex with sensory, motor, and association cortices. *The Journal of Comparative Neurology, 226,* 184-202.

Miller, R. (1991). *Cortico-hippocampal interplay. Studies of brain functions. Vol. 17.* Berlin: Springer-Verlag.

Mishkin, M., Ungerleider, L.G., & Macko, K.A. (1983). Object vision and spatial vision: two cortical pathways. *Trends in Neurosciences, 6,* 414-417.

Misslin, R. (1983). *Contribution neuroéthologique à l'étude des conduites néotiques chez la souris.* Doctoral Dissertation, Université de Strasbourg.

Mitchell, S.J., & Ranck, J.B. (1980). Generation of theta rhythm in medial entorhinal cortex of freely moving rats. *Brain Research, 189,* 49-66.

Mittelstaedt, M.L., & Mittelstaedt, H. (1980). Homing by path integration in a mammal. *Naturwissenschaft, 67,* 566.

Mize, R.R., Wetzel, A.B., & Thompson, V.E. (1971). Contour discrimination in the rat following removal of posterior neocortex. *Physiology and Behavior, 6,* 241-246.

Mizumori, S.J.Y., & Williams, J.D. (1993). Directionally selective mnemonic properties of neurons in the lateral dorsal nucleus of the thalamus of rats. *Journal of Neuroscience, 16,* 4015-4028.

Mizumori, S.J.Y., McNaughton, B.L., & Barnes, C.A. (1989). Preserved spatial coding in hippocampal CA1 pyramides during reversible suppression of CA3c output: evidence for pattern completion in the hippocampus. *Journal of Neuroscience, 9,* 3915-3928.

Morris, R.G.M. (1981). Spatial localization does not require the presence of local cues. *Learning and Motivation, 12,* 239-260.

Morris, R.G.M., Barnard, A., & Biegler, R. (1994). Ibotanate hippocampal lesions impair spatial learning of a new landmark array without affecting retention of a familiar array. *European Journal of Neuroscience, Suppl. 7,* 6508.

Morris, R.G.M., Garrud, P., Rawlins, J.N.P., & O'Keefe, J. (1982). Place navigation in rats with hippocampal lesions. *Nature, 297,* 681-683.

Morris, R.G.M., Hagan, J.J., & Rawlins, J.N.P. (1986). Allocentric spatial learning by hippocampectomised rats: a further test of "spatial mapping" and "working memory" theories of hippocampal function. *Quarterly Journal of Experimental Psychology, 38B,* 365-395.

Morris, R.G.M., Schenk, F., Tweedie, F., & Jarrard, L.E. (1990). Ibotenate lesions of hippocampus and/or subiculum: dissociating components of allocentric spatial learning. *European Journal of Neuroscience, 2,* 1016-1028.

Muller, R.U., & Kubie, J.L. (1987). The effects of changes in the environment on the spatial firing of hippocampal complex-spike cells. *Journal of Neuroscience, 7,* 1951-1968.

Muller, R.U., Kubie, J.L. (1989). The firing of hippocampal place-cells predicts the future position of freely moving rats. *Journal of Neuroscience, 9,* 4101-4110.

Muller, R.U., Kubie, J.L., Bostock, E.M., Taube, J.S., & Quirk, G.J. (1991b). Spatial firing correlates of neurons in the hippocampal formation of freely moving rats. In J. Paillard (Ed.), *Brain and space* (pp. 296-333). Oxford: Oxford University Press.

Muller, R., Kubie, J.L., & Ranck, J. (1987). Spatial firing patterns for hippocampal complex-spike cells in a fixed environment. *Journal of Neuroscience, 7,* 1935-1950.

Muller, R.U., Kubie, J.L., & Saypoff, R. (1991a). The hippocampus as a cognitive graph (abridged version). *Hippocampus, 1,* 243-246.

Munn, N.L. (1950). *Handbook of psychological research on the rat.* Boston, MA: Houghton-Mifflin.

Nadel, L. (1991). The hippocampus and space revisited. *Hippocampus, 1,* 221-229.

Neisser, U. (1967). *Cognitive psychology*. New York: Appleton-Century-Crofts.

Neisser, U. (1976). *Cognition and reality.* San Francisco, CA: W.H. Freeman and Company.

O'Keefe, J. (1976). Place units in the hippocampus of the freely-moving rat. *Experimental Neurology, 51,* 78-109.

O'Keefe, J. (1988). Computations the hippocampus might perform. In L. Nadel, L.A. Cooper, P. Culicover, & R.M. Harmish (Eds.), *Neural connections and mental computations* (pp. 225-284). Cambridge, MA: MIT Press.

O'Keefe, J. (1990). A computational theory of the hippocampal cognitive map. In J. Storm-Mathisen, J. Zimmer, & O.P. Ottersen (Eds.), *Progress in brain research* (Vol. 83, pp. 301-312). Amsterdam: Elsevier.

O'Keefe, J. (1991a). An allocentric spatial model for the hippocampal cognitive map. *Hippocampus, 1,* 230-235.

O'Keefe, J. (1991b). The hippocampal cognitive map and navigational strategies. In J. Paillard (Ed.), *Brain and space* (pp. 273-295). Oxford: Oxford University Press.

O'Keefe, J., & Burgess, N. (1996). Geometric determinants of the place fields of hippocampal neurons. *Nature* (in press)

O'Keefe, J., & Conway, D.H. (1978). Hippocampal place units in freely moving rat: why they fire where they fire. *Experimental Brain Research, 31,* 573-590.

O'Keefe, J., & Conway, D.H. (1980). On the trail of the hippocampal engram. *Physiological Psychology, 8,* 229-238.

O'Keefe, J., & Dostrovsky, J. (1971). The hippocamus as a spatial map. Preliminary evidence from unit activity in the freely-moving rat. *Brain Research, 34,* 171-175.

O'Keefe, J., & Nadel, L. (1978). *The hippocampus as a cognitive map.* London: Oxford University Press.

O'Keefe, J., & Recce, M. (1993). Phase relationship between hippocampal place units and the EEG theta rhythm. *Hippocampus, 3,* 317-330.

O'Keefe, J., & Speakman, A. (1987). Single unit activity in the rat hippocampus during a spatial memory task. *Experimental Brain Research, 68,* 1-27.

O'Mara, S.M. (1995). Spatially selective firing properties of hippocampal formation neurons in rodents and primates. *Progress in Neurobiology, 45,* 235-274.

Olton, D.S. (1978). Characteristics of spatial memory. In S.H. Hulse, H. Fowler, & W.K. Honig (Eds.), *Cognitive processes in animal behavior* (pp. 341-373). Hillsdale, NJ: Lawrence Erlbaum Associates.

Olton, D.S. (1979). Mazes, maps, and memory. *American Psychologist, 34,* 583-596.

Olton, D.S. (1982). Spatially organized behaviors of animals: behavioral and neurological studies. In M. Potegal (Ed.), *Spatial abilities. Developmental and physiological foundations* (pp. 335-360). New York: Academic Press.

Olton, D.S., Becker, J.T., & Handelmann, G.E. (1980). Hippocampal function: working memory or cognitive mapping? *Physiological Psychology, 8,* 239-246.

Olton, D.S., Branch, M., & Best, P. (1978). Spatial correlates of hippocampal unit activity. *Experimental Neurology, 58,* 387-409.

Olton, D.S., & Collison, C. (1979). Intramaze cues and "odor trials" fail to direct choice behavior on an elevated maze. *Animal Learning and Behavior, 7,* 221-223.

Olton, D.S., Collison, C., & Werz, M.A. (1977). Spatial memory and radial arm maze performance of rats. *Learning and Motivation, 8,* 389-314.

Olson, C.R., & Graybiel, A.M. (1970). Sensory maps in the claustrum of the cat. *Nature, 288,* 479-481.

Olton, D.S., & Samuelson, R.J. (1976). Remembrance of places passed: spatial memory in rats. *Journal of Experimental Psychology: Animal Behavior Processes, 2,* 97-116.

Olton, D.S., Wenk, G.L., Church, R.M., & Meck, W.H. (1988). Attention and the frontal cortex as examined by simultaneous temporal processing. *Neuropsychologia, 26,* 307-318.

Pavlides, C., & Winson, J. (1989). Influences of hippocampal place cell firing in the awake state on the activity of these cells during subsequent sleep episodes. *Journal of Neuroscience, 9,* 2907-2918.

Paylor, R., Baskall, L., & Wehner, J.M. (1993). Behavioral dissociations between C57BL/6 and DBA/2 mice on learning and memory tasks: a hippocampal dysfunction hypothesis. *Psychobiology, 21,* 11-26.

Peters, A. (1896). Ueber die Beziehungen zwischen Orientierungstörungen und doppelseitiger Hemianopsie. *Archiv für Augenheilkunde, 32,* 175-187.

Piaget, J. (1937). *La construction du réel chez l'enfant.* Paris: Delachaux & Niestlé.

Piaget, J., & Inhelder, B. (1956). *The child's conception of space.* London: Routledge and Kegan Paul.

Pico, R.M., & Davis, J.L. (1984). The radial maze performance of mice: assessing the dimensional requirements for serial order memory in animals. *Behavioral and Neural Biology, 40,* 5-26.

Poucet, B. (1985). Spatial behaviour of cats in cue-controlled environments. *Quarterly Journal of Experimental Psychology, 37B,* 155-179.

Poucet, B. (1989). Object exploration, habituation and response to a spatial change following septal or medial frontal cortical damage. *Behavioral Neuroscience, 103,* 1009-1016.

Poucet, B. (1990). A further characterization of the spatial problem-solving deficit induced by lesions of the medial frontal cortex in the rat. *Behavioural Brain Research, 41,* 229-237.

Poucet, B. (1993). Spatial cognitive maps in animals: new hypotheses on their structure and neural mechanisms. *Psychological Review, 100,* 163-182.

Poucet, B., & Benhamou, S. (1996). The neuropsychology of spatial cognition in the rat. *Critical Reviews in Neurobiology* (in press).

Poucet, B., Bolson, B., & Herrmann, T. (1990). Spatial behaviour of normal and septal rats on alternate route maze problems. *Quarterly Journal of Experimental Psychology, 42B,* 369-384.

Poucet, B., & Buhot, M.-C. (1989). Scopolamine impairs response-to-change based on distal cue in the rat. *Physiology and Behavior, 46,* 355-359.

Poucet, B., & Buhot, M.-C. (1994). Effects of medial septal or unilateral hippocampal inactivations on reference and working spatial memory in rats. *Hippocampus, 4,* 315-321.

Poucet, B., Buhot-Averseng, M.-C., & Thinus-Blanc, C. (1983a). Food-searching behavior of cats in a multiple-choice elimination problem. *Learning and Motivation, 14,* 140-153.

Poucet, B., Chapuis, N., Durup, M., & Thinus-Blanc, C. (1986). Exploratory behavior as an index of spatial knowledge in hamsters. *Animal Learning and Behavior, 14,* 93-100.

Poucet, B., Cressant, A., & Muller, R.U. (1995). The positional firing of place cells is controlled more strongly by cards on the wall than by 3-dimensional objects inside the arena. *Society for Neuroscience Abstracts, 21*, part 2, 946.

Poucet, B., Durup, M., & Thinus-Blanc, C. (1988). Activité exploratoire et connaissance de l'espace chez le Hamster doré (*Mesocricetus auratus*); comparaison avec d'autres espèces (*Meriones unguiculatus, Rattus norvegicus*). *Sciences et Techniques des Animaux de Laboratoire, 13,* 135-138.

Poucet, B., & Herrmann, T. (1990). Septum and medial frontal cortex contribution to spatial problem solving. *Behavioural Brain Research, 37,* 269-280.

Poucet, B., Herrmann, T., & Buhot, M.-C. (1991a). Effects of short-lasting inactivations of the ventral hippocampus and septum on long-term and short-term acquisition of spatial information in rats. *Behavioural Brain Research, 44,* 53-65.

Poucet, B., Lucchessi, H., & Thinus-Blanc, C. (1991b). What information is used by rats to update choices on the radial arm maze? *Behavioural Processes, 25,* 15-26.

Poucet, B., & Scotto, G. (1987). Memory properties of spatial behaviours in cats and hamsters. In P. Ellen & C. Thinus-Blanc (Eds.), *Cognitive processes and spatial orientation in animal and man* (pp. 135-146). Dordrecht, Neth.: Martinus-Nijhoff.

Poucet, B., & Thinus-Blanc, C. (1992). Cognition spatiale et organisation nerveuse. *Psychologie Française, 37,* 47-54.

Poucet, B., Thinus-Blanc, C., & Chapuis, N. (1983b). Route-planning in cats related to the visibility of the goal. *Animal Behaviour, 31,* 594-599.

Poucet, B., Thinus-Blanc, C., & Muller, R. (1993). A demonstration of place-cells in the ventral hippocampus in rats. *Society for Neuroscience Abstracts, 19,* part 1, 796.

Poucet, B., Thinus-Blanc, C., & Muller, R.U. (1994). Place cells in the ventral hippocampus of rats. *NeuroReport, 5,* 2045-2048.

Prescott, T.J. (1996). Spatial representation for navigation in animats. *Adaptive Behavior* (in press).

Quirk, G.J., Muller, R.U., & Kubie, J.L. (1990). The firing of hippocampal place-cells in the dark depends on the rat's recent experience. *Journal of Neuroscience,* 2008-2017.

Quirk, G.J., Muller, R.U., Kubie, J.L., & Ranck, J.B. Jr. (1992). The positional firing properties of medial entorhinal neurons: description and comparison with hippocampal place cells. *Journal of Neuroscience, 12,* 1945-1963.

Ranck, J.B. Jr. (1973). Studies on single neurons in dorsal hippocampal formation and septum in unrestrained rats. *Experimental Neurology, 41,* 461-555.

Ranck, J.B. Jr. (1984). Head-direction cells in the deep layers of dorsal presubiculum in freely moving rats. *Society for Neuroscience Abstracts, 10,* 599.

Ranck, J.B. Jr. (1985). Head directgion cells in the deep cell layer of dorsal presubiculum in freely moving rats. In G. Buzsaki & C.H. Vanderwolf (Eds.), *Electrical activity of the archicortex* (pp. 217-220). Budapest: Publishing House of the Hungarian Academy of Sciences.

Rashotte, M.E. (1987). Behavior in relation to objects in space: some historical perspectives. In P. Ellen & C. Thinus-Blanc (Eds.), *Cognitive processes and spatial orientation in animal and man* (pp. 39-54). Dordrecht, Neth.: Martinus-Nijhoff.

Rasmussen, M., Barnes, C.A., & McNaughton, B.L. (1989). A systematic test of cognitive mapping, working memory, and temporal discontinuity theories of hippocampal formation. *Psychobiology, 17,* 335-348.

Reep, R.L., Goodwin, G.S., & Corwin, J.V. (1990). Topographic organisation in the cortical connections of medial agranular cortex in rats. *Journal of Comparative Neurology, 294,* 262-280.

Regolin, L., Vallortigara, G., & Zanforlin, M. (1994). Perceptual and motivational aspects of detour behaviour in young chicks. *Animal Behaviour, 47,* 123-131.

Renner, M.J. (1988). Learning during exploration: the role of behavioral topography during exploration in determining subsequent adaptive behavior. *International Journal of Comparative Psychology, 2,* 43-56.

Restle, F. (1957). Discrimination of cues in mazes: a resolution of the "place vs response" question. *Psychological Review, 64,* 217-228.

Rieser, J.J., Guth, D.A., & Hill, E.W. (1986). Sensitivity to perspective structure while walking without vision. *Perception, 15,* 173-188.

Ritchie, B.F. (1948). Studies in spatial learning: VI. Place orientation and direction orientation. *Journal of Experimental Psychology, 38,* 658-669.

Ritchie, B.F., Aeschliman, B., & Pierce, P. (1950). Studies in spatial learning: VIII. Place performance and the acquisition of place dispositions. *Journal of Comparative and Physiological Psychology, 43,* 73-85.

Ritchie, B.F., Hay, A., & Hare, R. (1951). Studies in spatial learning. IX: A dispositional analysis of response performance. *Journal of Comparative and Physiological Psychology, 43,* 73-85.

Roberts, W.A. (1984). Some issues in animal spatial memory. In H.L. Roitblat, T.G. Bever, & H.S. Terrace (Eds.), *Animal cognition* (pp. 425-443). Hillsdale, NJ: Lawrence Erlbaum Associates.

Roitblat, H.L., Tham, W., & Golub, L. (1982). Performance of *Betta splendens* in a radial arm maze. *Animal Learning and Behavior, 10,* 108-114.

Roland, P.E., & Friberg, L. (1985). Localization of cortical areas activated by thinking. *Journal of Neurophysiology, 53,* 1219-1243.

Rossi-Arnaud, C., & Ammassari-Teule, M. (1994). Radial maze performance in inbred mice: evidence for strain-dependent neural nets subserving spatial abilities. *Psychobiology, 22,* 320-327.

Rossi-Arnaud, C., Fagioli, S., & Ammassari-Teule, M. (1991). Spatial learning in two inbred strains of mice: genotype-dependent effect of amygdaloid and hippocampal lesions. *Behavioural Brain Research, 42,* 25-32.

Roullet, P., & Lassalle, J.M. (1990). Genetic variation, hippocampal mossy fibers distribution, novelty reactions and spatial representation in mice. *Behavioural Brain Research, 41,* 61-69.

Rudy, J.W., & Sutherland, R.J. (1995). Configurational association theory and the hippocampal formation: an appraisal and reconfiguration. *Hippocampus, 5,* 375-389.

Rudy, J.W., Stadler-Morris, S., & Albert, P. (1987). Ontogeny of spatial navigation behaviors in the rat: dissociation of "proximal" and "distal"-cue-based behaviors. *Behavioral Neuroscience, 101,* 62-73.

Rumelhart, D.E., & Zipser, D. (1986). Feature discovery by competitive learning. In D.E. Rumelhart, J.M. McClelland and the PDP Research Group (Eds.), *Parallel distributed processing: Explorations in te microstructure of cognition* (Vol. 1, pp. 151-193). Cambridge, MA: The MIT Press.

Sakurai, Y., & Sugimoto, S. (1986). Multiple unit activity of prefrontal cortex and dorsomedial thalamus during delayed go/no-go alternation in the rat. *Behavioural Brain Research, 20,* 295-301.

Save, E. (1993). *Rôle du cortex pariétal associatif dans l'élaboration et l'utilisation des représentations spatiales chez le rat. Comparaison avec l'hippocampe.* Doctoral Dissertation, Université d'Aix-Marseille III.

Save, E. (1996). Interaction between the mapping and dead reckoning systems during place navigation learning in the rat (submitted).

Save, E., Buhot, M.C., Foreman, N., & Thinus-Blanc, C. (1992a). Exploratory activity and response to a spatial change in rats with hippocampal or posterior parietal cortical lesions. *Behavioural Brain Research, 47,* 113-127.

Save, E., Cressant, A., Thinus-Blanc, C., & Poucet, B. (1996a). Early visual deprivation does not prevent hippocampal place cell firing in the rat. *Society for Neuroscience Abstracts* (in press).

Save, E., Granon, S., Buhot, M.-C., & Thinus-Blanc, C. (1996b). Effects of limitations on the use of some visual and kinesthetic information in spatial mapping during exploration in the rat. *Quarterly Journal of Experimental Psychology.*

Save, E., & Moghaddam, M. (1996). Effects of lesions of the associative parietal cortex in the acquisition and use of spatial memory in egocentric and allocentric navigation tasks in the rat. *Behavioral Neuroscience, 110,* 75-85.

Save, E., Poucet, B., Foreman, N., & Buhot, M.C. (1992b). Object exploration, and reactions to spatial and non spatial changes in hooded rats following damage to parietal cortex or dorsal hippocampus. *Behavioral Neuroscience, 106,* 447-456.

Scardigli, P., Fabre, J.-C., & Thinus-Blanc, C. (1990). The effects of superior colliculus lesions on reactions to novelty in the hamster. *Behavioural Brain Research, 36,* 27-32.

Schenk, F. (1996). The Morris water maze (is not a maze). In N. Foreman & R. Gillett (Eds.), *A handbook of spatial research paradigms and methodologies*. Hove, UK: Lawrence Erlbaum Associates.

Schenk, F., Contant, B., & Grobéty, M.-C. (1990). Angle and directionality affect rat's organization of visit sequences and spatial learning in modular mazes. *Learning and Motivation, 21,* 164-189.

Schenk, F., Grobéty, M.-C., Lavenex, P., & Lipp, H.-P. (1995). Dissociation between basic components of spatial memory in rats. In E. Alleva et al. (Eds.), *Behavioural brain research in naturalistic and semi-naturalistic settings* (pp. 277-300). Kluwer Academic Press.

Schenk, F., Inglin, F., & Gyger, M. (1983). Activity and exploratory behavior after lesions of the medial entorhinal cortex in the woodmouse (*Apodemus sylvaticus*). *Behavioral and Neural Biology, 37,* 89-107.

Schenk, F., & Morris, R.G.M. (1985). Dissociation between components of spatial memory in rats after recovery from the effects of retrohippocampal lesions. *Experimental Brain Research, 58,* 11-28.

Schiller, P.H. (1949). Analysis of detour behavior. Learning of roundabout pathways in fish. *Journal of Comparative Physiological Psychology, 42,* 463-475.

Schmajuk, N.A., & Thieme, A.D. (1992). Purposive behavior and cognitive mapping: a neural network model. *Biological Cybernetics, 67,* 165-174.

Schneider, G.E. (1966). *Superior colliculus and visual cortex: contrasting behavioral effects of their ablation in the hamster.* PhD. Dissertation, Massachusett Institute of Technology

Schneider, G.E. (1969). Two visual systems: Brain mechanisms for localization and discrimination are dissociated by tectal and cortical lesions. *Science, 163,* 895-902

Schölkopf, B., & Mallot, H.A. (1995). View-based cognitive mapping and planning. *Adaptive Behaviour, 3,* 311-348.

Scoville, W.B., & Milner, B. (1957). Loss of recent memory after bilateral hippocampal lesions. *Journal of Neurology and Psychiatry, 20,* 11-21.

Sharp, P.E. (1991). Computer simulation of hippocampal place cells. *Psychobiology, 19,* 103-115.

Sharp, P.E., Blair, H.T., Etkin, D.E., & Tzanetos, D.B. (1995). Influences of vestibular and visual motion information on the spatial firing patterns of hippocampal place cells. *The Journal of Neuroscience, 15,* 173-189.

Sharp, P.E., McNaughton, B.L., & Barnes, C.A. (1985). Enhancement of hippocampal field potentials in rats exposed to a novel, complex environment. *Brain Research, 339,* 361-365.

Shepard, J.F. (1931). More learning. *Psychological Bulletin, 28,* 240-241.

Shepard, J.F. (1933). Higher processes in the behavior of rats. *Proceedings of the National Academy of Sciences, 19,* 149-152.

Shepard, R.N. (1975). Form, formation, and transformation of internal representations. In R. Solso (Ed.), *Information processing and cognition: the Loyola Symposium* (pp. 87-122). Hillsdale, NJ: Lawrence Erlbaum Associates.

Shepard, R.N., & Podgorny, P. (1978). Cognitive processes that resemble perceptual processes. In W.K. Estes (Ed.), *Handbook of learning and cognitive processes* (pp. 189-237). Hillsdale, NJ: Lawrence Erlbaum Associates.

Sherry, D.F. (1982). Food storage, memory, and marsh tits. *Animal Behaviour, 30,* 631-633.

Sherry, D.F. (1984). Food storage by black-capped chickadees: memory for the location and contents of caches. *Animal Behaviour, 32,* 451-466.

Sherry, D.F. (1987). Spatial memory in food-storing birds. In P. Ellen & C. Thinus-Blanc (Eds.), *Cognitive processes and spatial orientation in animal and man* (Vol. I, pp. 305-321). Dordrecht, Neth.: Martinus Nijhoff.

Sherry, D.F., Krebs, J.R., & Cowie, R.J. (1981). Memory for the location of stored food in marsh tits. *Animal Behaviour, 29,* 1260-1266.

Shettelworth, S.J., & Krebs, J.R. (1986). Stored and encountered seeds: a comparison of two spatial memory tasks in marsh tits and chickadees.

Journal of Experimental Psychology: Animal Behavior Processes, 12, 248-257.

Soffié, M., Bronchart, M., & Lebailly, B. (1986). Scopolamine-induced deficits in acquisition of a complex spatial learning. *Physiology and Behavior, 37,* 79-84.

Sokolov, E.N. (1960). Neuronal models and the orienting reflex. In M.A.B. Brazier (Ed.), *The central nervous system* (pp. 187-276). New York: Josiah Macy Jr. Foundation.

Sokolov, E.N. (1963a). *Perception and the conditioned reflex.* (Transl. S.W. Waydenfield.). Oxford: Pergamon Press.

Sokolov, E.N. (1963b). Higher nervous functions: the orienting reflex. *Annual Review of Physiology, 25,* 545-580.

Spence, K.W., & Lippitt, R. (1946). An experimental test of the sign-Gestalt theory of trial and error learning. *Journal of Experimental Psychology, 36,* 491-502.

Stein, J.F. (1991). Space and the parietal association areas. In J. Paillard (Ed.), *Brain and space* (pp. 185-222). Oxford: Oxford University Press.

Stephens, B., & Krebs, J. (1986). *Foraging theory.* Princeton, NJ: Princeton University Press.

Steward, O. (1976). Topographic organization of the projections from the entorhinal area of the hippocampal formation of the rat. *Journal of Comparative Neurology, 167,* 285-314.

Strauss, M.S. (1979). Abstraction of prototypical information by adults and 10 months-old infants. *Journal of Experimental Psychology: Human Learning and Memory, 5,* 618-635.

Strelow, E.R. (1985). What is needed for a theory of mobility: direct perception and cognitive maps. Lessons from the blind. *Psychological Review, 92,* 226-248.

Sutherland, N.S. (1957). Spontaneous alternation and stimulus avoidance. *Journal of Comparative and Physiological Psychology, 50,* 358-362.

Sutherland, R.J., Arnold, K.A., & Rodriguez, A.R. (1987). Anterograde effects on place memory after limbic or diencephalic damage. *Society for Neuroscience Abstracts, 15,* 607.

Sutherland, R.J., Chew, G.L., Baker, J.C., & Linggard, R.C. (1987). Some limitations on the use of distal cues in place navigation by rats. *Psychobiology, 15,* 48-57.

Sutherland, R.J., & Dyck, R.H. (1984). Place navigation by rats in a swimming pool. *Canadian Journal of Psychology, 38,* 322-347.

Sutherland, R.J., Kolb, B., & Whishaw, I.Q. (1982). Spatial mapping: definitive disruption by hippocampal or medial frontal cortical damage in the rat. *Neuroscience Letters, 31,* 271-276.

Sutherland, R.J., & Linggard, R.C. (1982). Being there: a novel demonstration of latent spatial learning in the rat. *Behavioral and Neural Biology, 32,* 103-107.

Sutherland, R.J., & Rudy, J.W. (1989). Configurational association theory: the role of the hippocampal formation in learning, memory and amnesia. *Psychobiology, 17,* 129-144.

Suzuki, S., Augerinos, G., & Black, A.H. (1980). Stimulus control of spatial behavior on the eight-arm maze in rats. *Learning and Motivation, 11,* 1-18.

Swanson, L.W. (1981). A direct projection from Ammon's horn to prefrontal cortex in the rat. *Brain Research, 217,* 150-154.

Taube, J.S. (1992). Qualitative analysis of head-direction cells recorded in the rat anterior thalamus. *Society for Neuroscience Abstracts, 18,* 708.

Taube, J.S., Muller, R.E., & Ranck, J.B.J. (1990a). Head direction cells recorded from the postsubiculum in freely moving rats. I. Description and quantitative analysis. *Journal of Neuroscience, 10,* 420-435.

Taube, J.S., Muller, R.E., & Ranck, J.B.J. (1990b). Head direction cells recorded from the postsubiculum in freely moving rats. II. Effects of environmental manipulations. *Journal of Neuroscience, 10,* 336-447.

Teroni, E., Portenier, V., & Etienne, A.S. (1987). Spatial orientation of the golden hamster in conditions of conflicting location-based and route-based information. *Behavioral Ecology and Sociobiology, 20,* 389-397.

Teule, M., Buhot, M.-C., & Durup, H. (1972). Apprentissage d'élimination progressive chez le Hamster doré (*Mesocricetus auratus*). *Psychologie Française, 17,* 175-182.

Teyler, T.J., & DiScenna, P. (1985). The role of hippocampus in memory: a hypothesis. *Neuroscience and Behavioral Review, 9,* 377-389.

Teyler, T.J., & DiScenna, P. (1986). The hippocampal memory indexing theory. *Behavioral Neuroscience, 100,* 147-154.

Thinus-Blanc, C. (1975). Données comportementales sur la discrimination visuelle chez l'animal. *L'Année Psychologique, 7,* 457-491.

Thinus-Blanc, C. (1978). Volume discrimination in golden hamsters. *Perceptual and Motor Skills, 46,* 1007-1015.

Thinus-Blanc, C. (1981). Volume discrimination learning in golden hamsters: effects of the structure of complex rearing cages. *Developmental Psychobiology, 14,* 397-403.

Thinus-Blanc, C. (1982). Selection of relevant cues in volume discrimination by golden hamsters reared in different environments. *Canadian Journal of Psychology, 36,* 520-526.

Thinus-Blanc, C. (1983). *De l'espace perçu à l'espace représenté: étude des mécanismes d'orientation spatiale chez les mammifères*. Doctoral Dissertation, Université d'Aix-Marseille II, 335 pp.

Thinus-Blanc, C. (1984). A propos des cartes cognitives chez l'animal: examen critique de l'hypothèse de Tolman. *Cahiers de Psychologie Cognitive, 4,* 537-558.

Thinus-Blanc, C. (1987). The concept of cognitive maps and its consequences. In P. Ellen & C. Thinus-Blanc (Eds.), *Cognitive processes and spatial orientation in animal and man* (Vol. 1, pp. 1-19). Dordrecht, Netherl.: Martinus-Nijhoff.

Thinus-Blanc, C. (1988a). Animal spatial cognition. In L. Weiskrantz (Ed.), *Thought without language* (pp. 371-395). Oxford: Oxford University Press.

Thinus-Blanc, C. (1988b). Exploration and memory. *Behavioral and Brain Sciences, 11,* 552-554.

Thinus-Blanc, C. (1992). Mémoirc spatiale distribuée. In *Percevoir, raisonner, agir. Articulations de modèles cognitifs* (Actes du 5e Colloque de l'ARC, pp. 145-157). Vandoeuvre: CRIN.

Thinus-Blanc, C. (1995). Animal spatial cognition. In H. Roitblatt & J.A. Meyer (Eds.), *Comparative approach to cognition* (pp. 326-345). Cambridge, MA: MIT Press.

Thinus-Blanc, C., Bouzouba, L., Chaix, K., Chapuis, N., Durup, M., & Poucet, B. (1987). A study of spatial parameters encoded during exploration in hamsters. *Journal of Experimental Psychology: Animal Behavior Processes, 13,* 418-427.

Thinus-Blanc, C., Durup, M., & Poucet, B. (1992). The spatial parameters encoded by hamsters during exploration: a further study. *Behavioural Processes, 26,* 43-57.

Thinus-Blanc C., & Foreman, N. (1993). A rapid test of rodent's vision using a modified open field apparatus. *Physiology and Behavior, 53,* 1003-1006.

Thinus-Blanc, C., & Gaunet, F. (1996). Space representations in the blind: Vision as a spatial sense? *Psychological Bulletin* (in press).

Thinus-Blanc, C., & Ingle, D. (1985). Spatial behaviour in gerbils. *Journal of Comparative Psychology, 99,* 311-315.

Thinus-Blanc, C., & Krimm, M. (1993). The effects of occipital lesions and duration of early visual deprivation on spatial behavior in the rat. *The European Journal of Neuroscience, suppl. 6,* 596.

Thinus-Blanc, C., & Lecas, J.-C. (1985). Effects of collicular lesions in the hamster during visual discrimination. An analysis from computer-video-actograms. *Quarterly Journal of Experimental Psychology, 37B,* 213-233.

Thinus-Blanc, C., Poucet, B., & Chapuis, N. (1982). Object permanence in cats: analysis in the space of locomotion. *Behavioural Processes, 7,* 81-86.

Thinus-Blanc, C., Save, E., Buhot, M.-C., & Poucet, B. (1991a). The hippocampus, exploratory activity and spatial memory. In J. Paillard (Ed.), *Brain and space*. Oxford University Press.

Thinus-Blanc, C., Save, E., & Poucet, B. (1996a). Animal spatial cognition and exploration. In N. Foreman & R. Gillett (Eds.), *Interacting with the environment: a handbook of spatial research paradigms and methodologies*. Mahwah, NJ: Erlbaum (in press).

Thinus-Blanc, C., Save, E. Poucet, B., & Buhot, M.-C. (1991b). The effects of reversible inactivation of hippocampus on exploratory activity and spatial memory. *Hippocampus, 1,* 363-369.

Thinus-Blanc, C., Save, E., Poucet, B., & Foreman, N. (1996b). Effects of parietal cortex lesions on spatial problem solving in the rat. *Behavioural Brain Research* (in press).

Thinus-Blanc, C., Save, E., Rossi-Arnaud, C., Tozzi, A., & Ammassari-Teule, M. (1996c). The differences shown by C57BL/6 and DBA/2 inbred mice in detecting spatial novelty are subserved by a different hippocampal and parietal cortex interplay. *Behavioural Brain Research* (in press).

Thinus-Blanc, C., & Scardigli, P. (1981). Object permanence in the golden hamster. *Perceptual and Motor Skills, 53,* 1010.

Thinus-Blanc, C., Scardigli, P., & Buhot, M.-C. (1991). The effect of superior colliculus lesions in hamsters: feature detection versus spatial localization. *Physiology and Behavior, 49,* 1-6.

Thorpe, W.H. (1963). *Learning and instinct in animals* (1rst ed., 1956). London: Methuen.

Toates, F. (1994a). What is cognitive and what is not cognitive? In D. Cliff, P. Husbands, & J.-A. Meyer (Eds.), *From animals to animats 3* (pp. 102-107). Cambridge, MA: The MIT Press.

Toates, F. (1994b). Hierarchies of control-changing weightings of levels. In M.-A. Rodrigues & M.-H. Lee (Eds.), *Perceptual control theory* (pp. 71-86). Aberystwyth: The University of Wales.

Toldi, J., Fehèr, O., & Wolff, J.R. (1986). Sensory interactive zones in the rat cerebral cortex. *Neuroscience, 18,* 461-466.

Tolman, E. (1932a). *Purposive behavior in animals and men.* New York: Appleton-Century-Crofts.

Tolman, E.C. (1932b). Lewin's concept of vectors. *Journal of General Psychology, 7,* 3-15.

Tolman, E.C. (1948). Cognitive maps in rats and men. *Psychological Bulletin, 55,* 189-208.

Tolman, E.C., & Honzik, C.H. (1930). Degrees of unger, reward and non-reward on maze learning in rat. *University of California Publications in Psychology, 4,* 241-256.

Tolman, E.C., Ritchie, B.F., & Kalish, D. (1946). Studies in spatial learning. II: Place learning versus response learning. *Journal of Experimental Psychology, 36,* 221-256.

Tomlinson, W.T., & Johnston, T.D. (1991). Hamsters remember spatial information derived from olfactory cues. *Animal Learning and Behavior, 19,* 185-190.

Torrealba, F., Olavarria, J., & Carrasco, M.A. (1984). Cortical connections of the anteromedial extrastriate visual cortex in the rat. *Experimental Brain Research, 56,* 543-549.

Touretzky, D.S., Redish, A.D., & Wan, H.S. (1993). Neural representation of space using sinusoidal arrays. *Neural Computation, 5,* 869-884.

Tsang, Y.-C. (1934). The functions of the visual areas of the cerebral cortex of the rat in the learning and retention of the maze. I. *Comparative Psychology Monographs, 10* (4, serial n° 50).

Tsang, Y.-C. (1936). The functions of the visual areas of the cerebral cortex of the rat in the learning and retention of the maze. II. *Comparative Psychology Monographs, 12* (2, serial n° 57).

Ungerleider, L.G., & Mishkin, M. (1982). Two cortical visual systems. In D.J. Ingle, M.A. Goodale, & R.J..W. Mansfield (Eds.), *Analysis of visual behavior*. Cambridge, MA: The MIT Press.

Upchurch, M., & Wehner, J. (1989). Inheritance of spatial learning abilities in inbred mice: a classical genetic analysis. *Behavioural Neuroscience, 103,* 1251-1258.

Uylings, H.B.M., & Van Eden, C.G. (1990). Qualitative and quantitative comparison of the prefrontal cortex in rat and in primates, including humans. In H.B.M. Uylings, C.G. van Eden, J.P.C. de Bruin, M.A. Corner, & M.G.P. Feenstra (Eds.), *The prefrontal cortex: its structure, function and pathology. Progress in brain research* (85, pp. 31-62). Amsterdam: Elsevier.

van Haaren, F., de Bruin, J.P.C., Heinsbroek, R.P.W., & van de Poll, N.E. (1985). Delayed spatial response alternation: effects of delay-interval duration and lesions of the medial prefrontal cortex on response accuracy of male and female Wistar rats. *Behavioural Brain Research, 18,* 41-49.

van Hoesen, G.W., & Pandya, D.N. (1975). Some connections of the entorhinal area (area 28) and perirhinal area 35) cortices in the monkey. I. Temporal lobe afferents. *Brain Research, 95,* 1-24.

Vanderwolf, C.H. (1969). Hippocampal electrical activity and voluntary movement in the rat. *Electroencephalography and Clinical Neurophysiology, 26,* 407-418.

Vauclair, J. (1990). Les images mentales chez l'animal. *La Recherche, 21,* 1006-1014.

von der Malsburg, C. (1973). Self-organizing of orientation sensitive cells in the striate cortex. *Kybernetik, 14,* 85-100.

Walker, J.A., & Olton, D.S. (1979). Spatial memory deficit following fimbria-fornix lesions: independent of time for stimulus processing. *Physiology and Behavior, 23,* 11-15.

Wehner, R., & Wehner, S. (1990). Insect navigation: use of maps or Ariadne thread? *Ethology, Ecology and Evolution, 2,* 27-48.

Whishaw, I.Q. (1987). Hippocampal, granule cell and CA3-4 lesions impair formation of a place learning-set in the rat and induce reflex epilepsy. *Behavioural Brain Research, 24,* 59-72.

Whishaw, I.Q. (1991). Latent learning in a swimming pool place task by rats: evidence for the use of associative and not cognitive mapping processes. *Quarterly Journal of Experimental Psychology, 43B,* 83-103.

Whishaw, I.Q., & Mittleman, G. (1986). Visits to starts, routes, and places by rats (*Rattus norvegicus*) in swimming pool navigation tasks. *Journal of Comparative Psychology, 100,* 422-431.

Whishaw, I.Q., & Vanderwolf, C.H. (1971). Hippocampal EEG and behaviour: effects of variation in body temperature and relation of EEG to vibrissae movement, swimming and shivering. *Physiology and Behavior, 6,* 391-397.

Whishaw, I.Q., & Vanderwolf, C.H. (1973). Hippocampal EEG and behaviour: changes in amplitude and frequency of RSA (theta rhythm) associated with spontaneous and learned movement patterns in rats and cats. *Behavioural Biology, 8,* 461-484.

Wiener, S.I. (1993). Spatial and behavioral correlates of striatal neurons in rats performing a self-initiated navigation task. *Journal of Neuroscience, 13,* 3802-3817.

Wiener, S.I., Korshunov, V.A., Garcia, R., & Berthoz, A. (1995). Inertial, substratal and landmark cue control of hippocampal CA1 place cell activity. *European Journal of Neuroscience, 7,* 2206-2219.

Wiener, S.I., Paul, C.A., & Eichenbaum, H. (1989). Spatial and behavioral correlates of hippocampal neural activity. *The Journal of Neuroscience, 9,* 2737-2763.

Wigstrom, H., Gustaffson, B. (1985). On long-lasting potentiation in the hippocampus: a proposed mechanism for its dependance on coincident pre- and post-synaptic activity. *Acta Physiologica Scandinavica, 123,* 519-522.

Wilkie, D.M., Spetch, M.L., & Chew, L. (1981). The ring dove's short-term memory capacity for spatial information. *Animal Behaviour, 29,* 639-641.

Wilson, M.A., & McNaughton, B.L. (1993). Dynamics of the hippocampal ensemble code for space. *Science, 261,* 1055-1058.

Wilson, M.A., & McNaughton, B.L. (1994). Reactivation of hippocampal ensemble memories during sleep. *Science, 265,* 676-679.

Wiltschko, W., & Balda, R.P. (1989). Sun compass orientation in seed-caching scrub jays (*Aphelocoma coerulescens*). *Journal of Comparative Physiology, 164,* 717-721.

Wilz, K.J., & Bolton, R.L. (1971). Exploratory behaviour in response to the spatial rearrangement of familiar stimuli. *Psychonomic Science, 24,* 117-118.

Worden, R. (1992). Navigation by fragment fitting: a theory of hippocampal function. *Hippocampus, 2,* 165-188.

Xavier, G.F., Saito, M.I., & Stein, C.S. (1991). Habituation of exploratory activity to new stimuli, to the absence of a previously presented stimulus and to new contexts, in rats. *Quarterly Journal of Experimental Psychology, 43B,* 157-173.

Xavier, G.F., Stein, C., & Bueno, O.F. (1990). Rats with dorsal hippocampal lesions do react to new stimuli but not to spatial changes of known stimuli. *Behavioral and Neural Biology,* 172-183.

Yoerg, S.I., & Kamil, A.C. (1982). Response strategies in the radial arm maze: running around in circles. *Animal Learning and Behaviour, 10,* 530-534.

Yoshino, R. (1991). A note on cognitive maps: an optimal spatial knowledge representation. *Journal of Mathematical Psychology, 35,* 371-393.

Zanforlin, M., & Poli, G. (1970). The burrowing rats: a new technique to study place learning and orientation. *Attice Memorie dell'Academia patavia di Scienze, Lettere e Arti, 82,* 653-670.

Zilles, K. (1990). Anatomy of the neocortex: cytoarchitecture and myeloarchitecture. In B. Kolb & R.C. Tees (Eds.), *The cerebral cortex of the rat*. Cambridge, MA: The MIT Press.

Zipser, D., & Andersen, R.A. (1988). A back-propagation programmed network that stimulates response properties of a subset of posterior parietal neurons. *Nature, 331,* 679-684.

Zoladek, L., & Roberts, W.A. (1978). The sensory basis of spatial memory in the rat. *Animal Learning and Behaviour, 6,* 77-81.

SUBJECT INDEX